EU food law

Related titles from Woodhead's food science, technology and nutrition list:

Food labelling (ISBN: 1 85573 496 6)

This unique new reference book provides a comprehensive guide to producing labelling that complies with existing legislation. The book deals with each aspect of food labelling from EU and US perspectives. The book covers major themes including the declaration of ingredients and additives, durability indication, nutrition and health information. Readers will find answers to questions raised by issues of GM and irradiated foods as well as guides to good practice.

Making the most of HACCP (ISBN: 1 85573 504 0)

Based on the experience of those who have implemented HACCP systems successfully, this book will meet the needs of food processing businesses at all stages of the HACCP system development. The collection is edited by two internationally recognised HACCP experts and includes chapters reflecting the experience of both major companies such as Cargill, Heinz and Sainsbury and the particular challenges facing SMEs. The scope of the book is truly international with experiences of HACCP implementation in countries such as Thailand, India, China and Poland.

Auditing in the food industry (ISBN: 185573 450 8)

Increasing consumer expectations, government legislation and levels of competition have led to a growing number of standards against which food processors are audited. This unique book provides food processors with a guide to the main types of audit, their importance, how to prepare for them and how to use them to gain competitive advantage. It will be an essential resource for all quality assurance and production managers within the food industry as well as auditors from both government and third-party agencies.

Details of these books and a complete list of Woodhead's food science, technology and nutrition titles can be obtained by:

- visiting our web site at www.woodhead-publishing.com
- contacting Customer services (e-mail: sales@woodhead-publishing.com; fax: +44 (0) 1223 893694; tel: + 44 (0) 1223 891358 ext. 30; address: Woodhead Publishing Limited, Abington Hall, Abington, Cambridge CB1 6AH, England)

If you would like to receive information on forthcoming titles in this area, please send your address details to: Francis Dodds (address, tel. and fax as above; e-mail: francisd@woodhead-publishing.com). Please confirm which subject areas you are interested in.

EU food law

A practical guide

Edited by
Kaarin Goodburn

CRC Press
Boca Raton Boston New York Washington, DC

WOODHEAD PUBLISHING LIMITED
Cambridge England

Published by Woodhead Publishing Limited
Abington Hall, Abington
Cambridge CB1 6AH
England
www.woodhead-publishing.com

Published in North and South America by CRC Press LLC
2000 Corporate Blvd, NW
Boca Raton FL 33431
USA

First published 2001, Woodhead Publishing Limited and CRC Press LLC

British Library Cataloguing in Publication Data
A catalogue record for this book is available from the British Library.

Library of Congress Cataloging-in-Publication Data
A catalog record for this book is available from the Library of Congress.

Woodhead Publishing Limited ISBN 1 85573 557 1
CRC Press ISBN 0-8493-1215-9
CRC Press order number: WP1215

Cover design by The ColourStudio
Project managed by Macfarlane Production Services, Markyate, Hertfordshire
(e-mail: macfarl@aol.com)
Typeset by MHL Typesetting Limited, Coventry, Warwickshire
Printed by TJ International, Padstow, Cornwall, England

Contents

Contributors

Chapters 1 and 2

Ms Kaarin Goodburn
Pinebridge
11 Yewfield Road
London NW10 9TD
England

Tel: +44 (0) 20 8451 0503
Fax: +44 (0) 20 8459 8061
E-mail: kgoodburn@bigfoot.com

Chapter 3

Mr Michael Fogden
Meat and Livestock Commission
PO Box 44
Winterhill House
Snowden Drive
Milton Keynes MK6 1AX
England

Tel: +44 (0) 1908 844177/677577
Fax: +44 (0) 1908 844302/609221
E-mail: michael_fogden@mlc.org.uk

Chapter 4

Ms D. W. Flowerdew
8 Broomhurst Court
Ridgeway Road
Dorking RH4 3EY
England

Tel/Fax: +44 (0) 1306 886626
E-mail:
d.flowerdew@foodlaw.demon.co.uk

Chapter 5

Dr S. Slorach
Swedish National Food
Administration
Box 622
75126 Uppsala
Sweden

E-mail: stsl@slv.se

Chapter 6

Ms Joan Sylvain Baughan
Keller and Heckman LLP
1001 G Street, NW
Suite 500W
Washington, DC 20001
USA
Tel: +1 202 434 4147
Fax: +1 202 434 4646
E-mail: baughan@khlaw.com

Mr Jean-Philippe Montfort
Keller and Heckman LLP
Rue Blanche 25
B-1060 Brussels
Belgium

Tel: 32 (2) 541 05 70
Fax: 32 (2) 541 05 80
E-mail: montfort@khlaw.be

Chapter 7

Ms Denise Love
50 St. Peter's Street
London N1 8JT
England

Tel: +44 (0) 020 7359 1049

Chapter 8

Ms Lynn Insall
Food and Drink Federation
6 Catherine Street
London WC2B 5JJ
England

Tel: +44 (0) 020 7836 2460
Fax: +44 (0) 020 7836 0580
E-mail: lynn.insall@fdf.org.uk

Chapters 9, 10 and 12

Mr Peter Berry Ottaway
Berry Ottaway & Associates Limited
1a Fields Yard
Plough Lane
Hereford HR4 0EL
England

Fax: +44 (0) 1432 270808
E-mail:
berry.ottaway@dial.pipex.com

Chapter 11

Mr Richard Fox
Pura Food Products Limited
Technical Development Centre
Crabtree Manorway South
Belverdere DA17 6BB
England

Tel: +44 (0) 208 320 9000
Fax: +44 (0) 208 320 9003
E-mail: richard.fox@pura.co.uk

1

Introduction

K. Goodburn, Chilled Food Association, London

1.1 The development of EU food law

EU food safety legislation has evolved over 30 years or so, reflecting a blend of scientific, social, political and economic factors. As a result, there has, at times, been little coherence in its development, resulting in over-complex and fragmented measures and lack of consistency. It is only recently that the EU has developed a clear policy framework for food law.

The Treaty of Rome (1957) made no reference to guiding principles for food legislation. The initial approach of the European Commission was to concentrate on the obligations set out in Article 3 of the Treaty to ensure the free movement of foodstuffs within the common market. Much of the food legislation in this period was developed by the Internal Market Directorate General (DGIII). A good example was the compositional directives which were adopted in the 1970s concerning honey, sugars, preserved milks, coffee extracts, fruit juices and nectars, jams, jellies and marmalades and chocolate and chocolate products. These addressed the problem of differing definitions of such products between Member States, establishing common compositional and quality requirements which would allow such products to be traded freely within the common market. The Commission also began to look at issues of food safety. As an example the Agriculture Directorate General (DGVI) developed a series of directives governing the safe production of particular groups of animal products with detailed rules on methods of production and quality control. Much of this early legislation was complex and prescriptive and was increasingly criticised for being inflexible and bureaucratic.

The Commission increasingly recognised the need both for a clear set of guiding principles in EU food legislation and for a more flexible approach. In

1985 it produced a Communication entitled ‘Completion of the Internal Market: Community Legislation on Foodstuffs’ setting out a number of objectives for food legislation. These objectives were to:

- protect public health
- provide consumers with information and protection in matters other than health and ensure fair trading (for example by setting appropriate standards for labelling to allow consumers to make informed choices about food products)
- provide for the adequate and necessary official controls of foodstuffs.

The explicit statement of public health and consumer interests as goals for EU food regulation was subsequently incorporated into the Single European Act of 1986 and the Maastricht Treaty of 1992. This new direction can be seen, for example, in the 1990 Directive on nutritional labelling for foodstuffs (Council Directive 90/496/EEC) which set out standard requirements and formatting for listing the nutritional content of a food.

The Commission also recognised the need for a less prescriptive and more flexible approach to food legislation. In response to the conclusions of the Edinburgh Summit of December 1992, the EC undertook several initiatives to simplify existing EU food legislation, including a reconsideration of whether certain items of legislation were necessary, and the removal of unnecessarily restrictive provisions from existing legislation. For example, in April 1994 the EC presented a proposal to amend Directive 89/398/EEC on foodstuffs for particular nutritional purposes (PARNUTS) in order to reduce the number of specific directives for particular categories of such foods from 8 to 4. The compositional directives of the 1970s covering such foodstuffs as chocolate and chocolate products were also simplified. As an alternative to its earlier emphasis on detailed and prescriptive legislation, the Commission also moved to a more flexible approach in some areas. An example of this new approach was the 1993 General Food Hygiene Directive (93/43/EEC). In contrast to the previous detailed rules covering specific foodstuffs the Directive established the Hazard Analysis and Critical Control Point (HACCP) system as a generic approach for all sectors of the food industry to adopt. Rather than relying on detailed requirements from the Commission, responsibility was given to individual food businesses to develop HACCP systems appropriate to their needs within a broad framework of minimum hygiene standards and requirements. The Directive also encouraged the development of guides on good hygiene practice by food businesses together with interested parties such as national food agencies and consumer groups.

1.2 The 2000 White Paper on the General Principles of Food Law

Although the Commission had made a number of important changes in its approach to food legislation, it recognised that it needed to do more if it was to meet the objectives set out in the mid-1980s, particularly, in such areas as food safety. During the 1990s in particular, consumer concerns about the content of their food, its method of manufacture, its safety and impact on their health increased. These concerns, and increasing criticism of EU food policy, were fuelled by such crises as that over BSE in meat, and other developments such as the use of genetically-modified organisms in food manufacture and the emergence of a new generation of 'functional' foods with major implications for nutrition and health. Such concerns lay behind the publication in May 1997 of the Commission's Green Paper on the General Principles of Food Law in the European Union (EC, 1997), which subsequently formed the basis for the White Paper of January 2000 (EC, 2000a). The White Paper set out a number of objectives:

- to improve the efficiency and coherence of EU food legislation, particularly in the area of food safety
- to restore consumer confidence by the above measures and by improving the quality of information available to consumers
- to extend the scope of EU food regulation by developing an EU-wide nutrition policy.

To achieve these objectives in the area of food safety, the White Paper set out a number of guiding principles, including:

- restating the overiding importance of consumer health and choice as the starting point for all EU food legislation
- adoption of the precautionary principle
- extending the scope of food safety regulation across the entire food chain from 'farm to fork' including, for example, relevant controls on animal feed
- the attribution of primary responsibility for safe food production to industry producers and suppliers within a framework set out by EU legislation
- setting out clear responsibilities for public bodies in setting standards for the food industry to meet and monitoring industry compliance
- establishing traceability as a major responsibility in food production as a prerequisite both to food safety and effective consumer choice
- the provision of independent scientific advice as a sound basis for legislation and consumer confidence
- establishing effective crisis management procedures, including an effective rapid alert system, to control food safety problems once they arise.

The White Paper set out some 80 separate actions and areas of work to implement these principles. These actions included plans for a new European Food Authority (EFA), designed to take over the EC's scientific work on food matters (EC, 2000b). The EFA will be an independent body which will report its findings directly to the public, as well as assuming responsibility for the EU's rapid alert system. Some of the specific tasks it may be required to provide advice on include:

- keeping the EU lists on authorised additives up to date and clarifying the status of enzymes
- research on the toxicological effects of substances naturally present in flavourings
- the development of proposals for directives on food supplements and fortified foods
- reviews of food contact materials and their labelling requirements, and
- the role of irradiation as a means of food preservation.

The reorganisation of responsibilities between the EC's various Directorate Generals in February 1997 also reflected the priorities of the White Paper and need for changes in the development of EU food regulation. Whereas much early EU food legislation had come from DGIII (Internal Market) and DGVI (Agriculture), responsibility for scientific advice and legislation governing consumer health was now given to DGXXIV (Health and Consumer Protection), known as DG SANCO. There was now a Directorate General formally responsible for monitoring and improving food safety and for the research and expertise on which the preparation of EU food law depends (until the EFA is ready to take over the Directorate General's work on scientific matters). Responsibility for the preparation of the relevant legislation remained with DGIII (Internal Market), DGV (Environment) and DGVI (Agriculture).

1.3 The structure of this book

As this very brief introduction has shown, the lack of an agreed initial strategy for EU food legislation has given rise to piecemeal development and a number of changes in direction. These beginnings have produced a sometimes inconsistent body of EU food law which can be very confusing to the lay person. This book seeks to cast some much-needed light on this complex and evolving body of law. It can, of course, only be a snap shot of a picture that is constantly changing. It is designed as a starting point for all those in the food chain, from producers to consumers, in grasping some of the key themes of EU food law. It tries to provide the foundation on which readers can build an effective understanding of the legislation that affects them.

An initial chapter introduces the key EU institutions and the legislative process as a foundation for the chapters on particular aspects of EU food law that follow. That chapter refers to the distinction between 'horizontal' legislation affecting all foodstuffs and 'vertical' legislation dealing with specific foods. Most of the chapters in this book deal with 'horizontal' themes. Part I considers various aspects of food safety from setting appropriate hygiene standards for the food industry to the control of additives in food, measures for avoiding contamination from such substances as pesticides, and the regulation of food contact materials such as packaging. Part II looks at ways EU food law has sought to provide consumers with the relevant information on which to make the right choice about the food they eat, for example through clear, accurate and consistent labelling. Finally, Part III contains two case studies which illustrate how these diverse 'horizontal' themes come together and impact 'vertically' on a particular foodstuff, looking first at an established ingredient in the manufacture of many foods, frying oil, and, secondly, at the emerging category of 'functional' foods which pose new challenges to EU food law makers.

1.4 References and further reading

EC, 1997, 'Green Paper on General Principles of Food Law in the EU', European Commission (COM (97)176 final).

EC, 2000a, 'White Paper on Food Safety', European Commission, (COM(2000)1).

EC, 2000b, 'Food law from farm to table – creating a European Food Authority', European Commission Press Release, DN: IP/00/1270, 8 August 2000.

EC, 2000c, European Commission Proposals to Consolidate and Simplify EU Legislation – various proposals, Commission of the European Communities, COM 2000 438 final.

2

EU institutions and the legislative process

K. Goodburn, Chilled Food Association, London

2.1 The EU and its institutions

The European Union (EU), at the time of writing, comprises fifteen member states with a population of approximately 370 million, and eleven 'official' languages (shown in Table 2.1). This picture is projected to change in the next few years as the EU has an ambitious programme of enlargement (see below). The EU is the result of the process of cooperation and integration which began in 1951, and is based on three treaties signed by the six founding members, Belgium, France, Germany, Italy, Luxembourg and the Netherlands:

1. Paris (1951), establishing the European Coal and Steel Community (ECSC).
2. Rome (1957), which established the European Economic Community (EEC).
3. Euratom (1957), which was signed in Rome and established the European Atomic Energy Community.

The main original objective of the European Community was the establishment of an internal market, an area without internal frontiers guaranteeing movement for goods, people, services and capital. There have been three major reforms of European law, introduced by further treaties, bringing about institutional changes and introducing new areas of responsibility:

1. The 1986 Single European Act (Luxembourg and The Hague) – came into force 1 June 1987.
2. The 1992 Treaty on European Union (Maastricht Treaty) – came into force 1 November 1993, established the EU and created the concept of European citizenship.
3. The 1999 Treaty of Amsterdam – came into force 1 May 1999.

Table 2.1 Member states and official languages of the EU

Member state	Official language
Austria	
Belgium	
Denmark	Danish
Finland	Finnish
France	French
Germany	German
Greece	Greek
Ireland	
Italy	Italian
Luxembourg	
The Netherlands	Dutch
Portugal	Portuguese
Spain	Spanish
Sweden	Swedish
United Kingdom	English

The EU's current main objectives are the following:

- to promote economic and social progress (e.g. establishment of the single market in 1993, and launch of the single currency in 1999)
- to assert the identity of the EU internationally
- to introduce European citizenship
- to develop an area of freedom, security and justice
- to maintain and build on established EU law.

In March 1998, the EU launched work that, if successful, will enlarge the EU by another thirteen countries: Bulgaria, Cyprus, the Czech Republic, Estonia, Hungary, Latvia, Malta, Poland, Romania, the Slovak Republic, Slovenia and Turkey.

The EU has enlarged four times since the establishment of the EEC in 1957:

1. 1973: Denmark, Ireland, UK
2. 1981: Greece
3. 1986: Portugal and Spain
4. 1995: Austria, Finland and Sweden.

The latest proposed enlargement would be the largest in terms of the number of countries, the geographical area involved (which would increase the size of the EU by 34%) and population (increasing by 105 million). On enlargement the weighting of member states will be modified, with the Commission comprising one national of each of the member states. The big countries (France, Germany, Italy, Spain and the UK) will effectively give up their second Commissioner, and the weighting of member states in the Council will be readjusted, to ensure that a decision taken by a majority of member states corresponds to a sufficient percentage of the EU's population. However, the number of MEPs will not exceed 700. Once negotiations with the various states are concluded, Accession Treaties will be submitted for ratification by both sides. The conditions that each applicant has to achieve were laid down by the Copenhagen European Council in June 1993:

- They must have stable institutions which guarantee democracy, the rule of law, human rights, and respect and protection for minorities.
- There must be a viable market economy capable of withstanding competition from the EU.
- They must be able to sign up to the objectives of the EU.

A White Paper published in 1995 listed the laws and regulations that applicants should enact in economic policy to prepare the ground for their future accession.

There are five main institutions involved in running the EU, as set out in Article 4 of the Treaty of Rome:

1. The European Commission (EC).
2. The Council of the EU.
3. The European Parliament (EP).
4. The Court of Justice.
5. The Court of Auditors.

In addition, ECOSOC (the Economic and Social Committee) often plays a significant role in the development of food legislation.

2.2 The European Commission

Article 155 of the EEC Treaty (the Treaty of Rome) describes the functions and powers of the EC in order to ensure the proper functioning and development of the common market. The EC shall:

- ensure that the provisions of the Treaty and the measures taken by the institutions thereto are applied
- formulate recommendations or deliver opinions on matters dealt with in the Treaty, if it expressly so provides or if the EC considers it necessary
- have its own power of decision and participate in the shaping of measures taken by the Council and the EP in the manner provided for in the Treaty
- exercise the powers conferred on it by the Council for the implementation of the rules laid down by the latter.

Following from the first point above, the EC, in theory at least, therefore has the power to check whether Community Acts and the national laws that are based on them are actually fully complied with in practical terms at local or regional levels. Articles in individual Directives provide that member states shall report to the EC on the implementation of Directives. The EC, in brief:

- initiates proposals for legislation

- is guardian of the Treaties and ensures that EU legislation is applied correctly by the member states
- manages and executes EU policies and international trade relations.

2.3 The Directorate-Generals

The EC has 16,000 staff, a large proportion of whom are involved in translation activities. The administrative services of the Commission consists of thirty-six Directorate-Generals (DGs) and equivalent departments, which are divided into Directorates, and Directorates into units. The DGs are headed by a Director-General who reports to a Commissioner, each of whom has political and operational responsibility for one or more DGs (Table 2.2). There are twenty Commissioners, often referred to as the Commission, who are appointed for a five-year term, which is the same term as the life of the EP, but starting six months later. Commissioners are obliged to be completely independent of their national governments and to act only in the interest of the EU. The president of the Commission is appointed by common accord of the governments of the member states, subject to approval by the EP. The governments of the member states, in conjunction with the president, then

Table 2.2 Directorate-Generals, their Commissioners and Director Generals

Directorate-General	Commissioners	Director-General
Regional Policy and Cohesion	Michel Barnier (F)	Guy Crauser
Internal Market	Frits Bolkenstein (NL)	John Mogg
Research	Philippe Busquin (B)	Jorma Routti
Health and Consumer Protection	David Byrne (Irl)	Robert Coleman
Employment and Social Affairs	Anna Diamantopolou (Gr)	Allan Larsson
Agriculture and Fisheries	Franz Fischler (A)	Manuel Silva Rodriguez
Trade	Pascal Lamy (F)	Hans-Friedrich Beseler
Enterprise and Information Society	Erkki Liikanen (Fin)	Fabio Colosanti
Competition	Mario Monti (I)	Alexander Schaub
Development and Humanitarian Aid	Paul Nielson (DK)	Philip Lowe
External Relations	Chris Patten (UK)	Guy Legras
Education and Culture	Viviane Reding (L)	Spyros Pappas
Budget	Michaele Schreyer (L)	Jean-Paul Mingasson
Economic and Monetary Affairs	Pedro Solbes Mira (Sp)	Giovanni Ravasio
Enlargement	Gunther Verheugen (D)	Eneko Landáburu Illarramendi
Justice and Home Affairs	Antonio Vitorino (P)	Adrian Fortescue
Environment	Margot Wallström (Sw)	James Currie

nominate other members, and the entire Commission is subject to the EP's vote of approval. The EP can, at any time, require the resignation of the Commission *en bloc*, as happened in 1999.

The Commission meets once a week to adopt proposals, finalise policy papers and take other decisions required of it. At its meetings, each item is presented by the Commissioner responsible for the policy sector in question. Decisions are taken when necessary by a majority vote; when a decision has been adopted, it becomes Commission policy and it has the full support of all Commissioners. In addition to the staff of their DGs, each Commissioner has his or her own private office or 'cabinet', which consists of six officials who serve as the bridge between the Commissioner and the DGs. The work of the Commission is coordinated by its Secretariat-General. The Commission's proposals, actions and decisions are scrutinised and judged by various EU institutions. For example, the Commission attends all sessions of the EP and must explain and justify its policies if required by its members. It must reply to written or oral questions by MEPs.

2.4 The Directorate-Generals and EU food law

The most important DGs in terms of food law are the following:

- Internal Market
- Agriculture and Fisheries
- Environment
- Health and Consumer Protection.

The mission of the Internal Market Directorate-General is to ensure that the European internal market functions effectively, particularly in eliminating unjustified barriers to the free movement of goods and services so that products legally marketed in one member state can be freely marketed in other member states. The development of the internal market was a major area of activity for the Commission in the 1970s and 1980s in particular, culminating in the establishment of the single market in 1993. The emphasis has now shifted to balancing the free movement of goods with other issues such as consumer health, although the Commission continues to exercise the power to bring member states before the European Court for setting up unjustified barriers to the free movement of goods and services.

The DG for Agriculture is responsible for the implementation of EU policy in the areas of agriculture and rural development, while the main task of the DG for Fisheries is currently the conservation and management of marine resources. Working with the DG for Health and Consumer Protection, the DG for Agriculture has put forward a number of measures to manage food safety problems in the agricultural sector such as BSE.

The role of the Environment DG has been enhanced since the 1999 Treaty of Amsterdam which enshrined the principle of sustainable developments as a central aim of the EU. It has proposed measures in such areas as nature conservation, pollution and waste management. Much of its work has been concerned with establishing voluntary schemes such as the Eco-Management and Audit Scheme (EMAS) and eco-labelling which allow businesses to audit the impact of their activities on the environment and achieve certification for environmentally friendly production operations. As consumer concern about environmental issues increases, the importance of such schemes will increase. There is also increasing regulation in this area which affects the food industry, notably the 1994 EU Directive on Packaging and Packaging Waste which requires proof that packaging materials are being recovered and recycled by businesses. The impact of this Directive increased significantly in 2000 when its scope was extended to include all organisations using over 50 tonnes of packaging per annum and with a turnover of £1 million or more. It is likely that such regulation will increase and the work of the DG will grow in importance.

However, the most important DG in the area of food law is the DG for Health and Consumer Protection. Its mission is to protect EU consumers' health, safety and economic interests. Its food safety activities cover the entire food chain, from animal and plant health to the labelling of food products. The DG provided the basis for the EC's White Paper on Food Safety in January 2000. Its responsibilities include the following:

- the assessment of possible risks to consumer health
- proposing and monitoring legislation in such areas of agriculture as veterinary care and animal feed which affect consumer health
- proposing and monitoring legislation on hygiene and safe practices in food processing and distribution (including the retail and catering sectors)
- inspections within and outside the EU to ensure that appropriate measures to meet food hygiene and safety standards are being implemented effectively
- management of the EU's Scientific Committees responsible for consumer health (see section 2.5).

The structure of the DG with its constituent Directorates (and their component units) is shown in Fig. 2.1. The work of the various Directorates in proposing legislation lies behind most of the chapters in the book and is not discussed here. The following discussion focuses on the monitoring and inspection work of the DG, handled primarily by the Food and Veterinary Directorate. Its role is to monitor and control how member states and countries outside the EU implement EU legislation on food safety, animal and plant health, and animal welfare. An important goal is to develop a harmonised approach to control and inspection activity throughout the food chain, based on audit rather than inspection. Inspections have resulted in infringement proceedings being taken against some member states and strengthening of import controls for products from non-EU countries where poor official control regimes were identified. Recent activities have included the following:

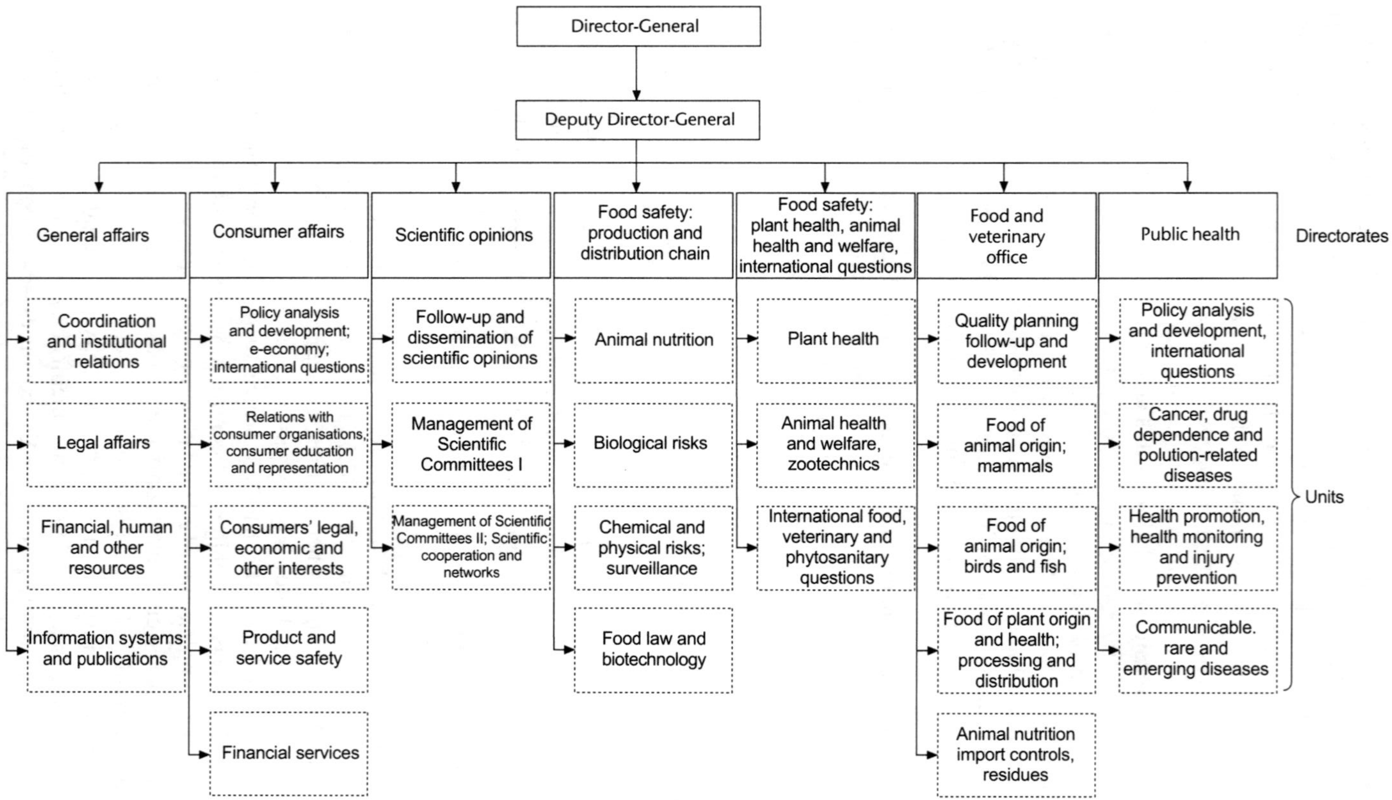

Fig. 2.1 The Directorate-General for Health and Consumer Protection.

- reviews of bovine and poultry meat production standards, looking, for example, at methods of approving abattoirs, hygiene standards and veterinary supervision of slaughter hygiene, and animal health
- inspections of procedures for the control of BSE, including cattle traceability, surveillance regimes, eradication plans, trade in animal waste products, the labelling and control of animal feed
- an assessment of border inspection posts within member states responsible for carrying out health checks on food of animal origin and animals from non-EU countries
- evaluation of national systems for the control of animal welfare including housing of animals, welfare at slaughter and long distance transport
- reviews of milk production in both member states and non-EU countries
- inspections to assess the control of contaminants, focusing on such areas as dioxins in animal feed, mycotoxins and pesticides
- specific inspections in response to a particular crisis, for example in reviewing arrangements for the control of dioxin contamination of animal products via animal feed in a number of member states after an incident of such contamination in Belgium in 1999.

The DG is responsible for the rapid alert system for foodstuffs presenting a direct risk to human health, set out as part of Directive 92/59/EEC on general product safety. Through the rapid alert system the DG has set up arrangements for collecting information on a crisis and circulating it to the competent authorities in each member state to minimise the risk to consumers. The rapid alert system was used, for example, in 1999 in response to the dioxin contamination incident in Belgium, leading to a rapid withdrawal of products that might be affected in a number of member states.

2.5 Scientific Committees

Several Directives and a number of Regulations provide for mandatory consultation of one or another of the Scientific Committees. A Commission Decision (EC, 1997) established a number of Scientific Committees within the EC, covering the following areas relating to consumer health and food safety:

- Scientific Committee for Food (SCF)
- Scientific Committee on Animal Nutrition
- Scientific Committee on Animal Health and Animal Welfare
- Scientific Committee on Veterinary Measures relating to Public Health
- Scientific Committee on Plants
- Scientific Committee on Cosmetic Products and Non-Food Products Intended for Consumers
- Scientific Committee on Medicinal Products and Medical Devices
- Scientific Committee on Toxicity, Ecotoxicity and the Environment.

Each Scientific Committee comprises no more than nineteen members, the number of members being determined by the EC in view of the expertise required. Members are scientific experts in one or more of the relevant fields of competence of the Committee in question. The Scientific Committees have an advisory role, being consulted where required by Community law and where the EC decides to consult them in relation to consumer health and food safety. In more detailed terms, the Scientific Committees' roles are as follows:

- to examine critically risk assessments made by scientists belonging to member state organisations
- to develop new risk assessment procedures relating to areas such as, for example, food-borne diseases and the transmissibility of animal diseases to humans
- to prepare scientific opinions designed to enable the EC to evaluate the scientific basis of the recommendations, standards and guidelines prepared in international forums
- to evaluate the scientific principles on which Community health standards are based, taking into account the risk assessment techniques developed by the international organisations concerned.

The Scientific Committees may draw the Commission's attention to any specific or emerging problem falling within their remit relating to consumer health and food safety. Working groups spanning several Scientific Committees may be established in order to prepare an opinion on the topic of concern, which may be required within a period set by the EC. The most important of these Scientific Committees in the area of food law is the Scientific Committee for Food (SCF), supported by a range of working groups. The SCF Working Group Structure is shown in Fig. 2.2.

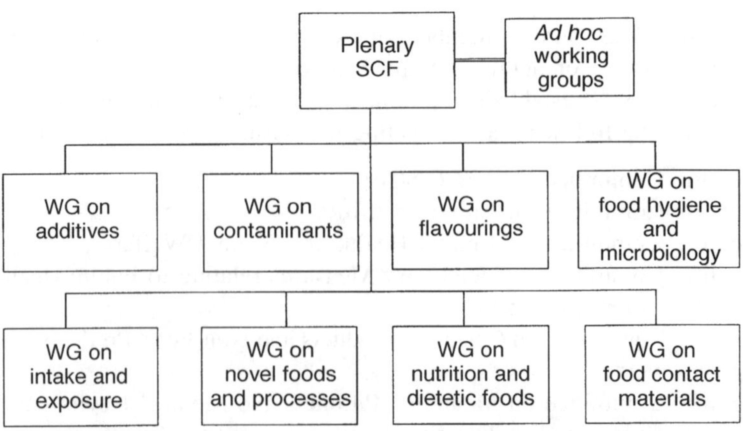

Fig. 2.2 SCF Working Group structure.

2.6 The European Food Authority

In November 2000 the EC proposed the creation of a new European Food Authority (EFA), designed to take over much of the advisory role currently undertaken by the Scientific Committees. Its role is designed to be advisory, both providing the EC with scientific risk assessments on all matters concerned with food safety, and assuming responsibility for communicating its findings direct to the public. It will also take over the rapid alert system. The current proposals are for the EFA to be run by a group of fourteen leading scientists who will be responsible for eight committees dealing with specific areas such as BSE, food additives and GMOs. The EFA will also liaise with member states via an advisory body made up of representatives from the competent authorities within the member states. The EFA will employ an estimated 300 people and have an annual budget of over £20 million.

2.7 The Council of the EU

The Council of the EU, usually known as the Council of Ministers, or Council, comprises fifteen member governments. Apart from the European Councils (summits), major decisions are taken at Foreign Minister level and at other times by appropriate ministers (Agriculture, Environment, etc). The Council exercises legislative and decision-making powers, modifying and/or approving proposals made by the EC. It is the forum in which the representatives of the fifteen member states can assert their interest and try to reach compromises. The Council is designed to ensure general coordination of the European Community's activities and is also responsible for intergovernmental cooperation, coordinating national policies, common foreign and security policy and in justice and home affairs.

The Council decides some matters by qualified majority voting (QMV), and others by unanimity. At the level of officials the Council is operated in Brussels by the Committee of Permanent Representatives (COREPER). The chairmanship (Presidency) of the Council rotates every six months, starting on 1 January and 1 July in accordance with a pre-established rota. Forthcoming presidencies are as follows:

- 2001: Sweden, Belgium
- 2002: Spain, Denmark
- 2003: Greece

The Presidency of the Council plays a vital part in the organisation of the work of the institution, particularly in driving the legislative and political decision-making process. The Presidency organises and chairs all Council meetings and seeks to work out compromises to resolve difficulties. Article 145 of the Treaty of Rome set out the role of the Council as being:

- to ensure coordination of the general economic policies of the member states
- to have power to take decisions

- to confer on the EC, in the Acts that the Council adopts, powers for the implementation of the rules that the Council lays down.

The Council may impose certain requirements in respect of the exercise of powers it confers on the EC, and may also reserve the right, in specific cases, to exercise directly implementing powers itself. The procedures referred to above must be consonant with principles and rules to be laid down in advance by the Council, acting unanimously on a proposal from the Commission and after obtaining the opinion of the EP.

Most provisions of the Treaty of Rome required a decision to be taken by the Council's unanimity. Nevertheless, some provisions provided for qualified majority. In addition, the Treaty foresaw the introduction of majority voting in many cases. Article 146 of the Treaty currently requires that, except when otherwise provided in the Treaty, the Council will act by a majority of its members. There are currently seventy-three articles and sub-articles subject to unanimous voting in the main EU Treaties. The bulk of these articles are within the areas of common foreign and security policy and police and judicial cooperation. However, even certain articles within the Community area are still covered by unanimity. These range from issues such as the appointment of the Council Secretary-General to issues like taxation. It is generally accepted that as the EU enlarges, its ability to take decisions by unanimity will become increasingly difficult. In order to ensure that the decision-making process will not grind to a halt, three options have been focused on:

1. A case-by-case approach.
2. Qualified majority voting as the general rule.
3. A categorisation of issues that could move to qualified majority voting.

In the Commission Opinion for the Intergovernmental Conference adopted on 26 January 2000, it was argued that qualified majority voting should become the general rule. The Commission identified what exceptions to that rule should be considered, listing five categories where it was possible to imagine unanimity being maintained in an enlarged Union of twenty-eight member states:

- Council decisions that have to be adopted by the member states in accordance with their constitutional requirements
- essential institutional decisions and those affecting the institutional balance (e.g. languages of the institutions)
- decisions in the fields of taxation and social security not related to the proper functioning of the internal market (e.g. harmonisation of legislation concerning certain forms of taxation)
- parallel internal and external decisions (e.g. association agreements)
- derogations from the common rules of the Treaty (e.g. compatibility of aid with the common market).

The Commission also subsequently adopted a specific contribution on the extension of QMV in the areas of taxation and social security. Where the

Table 2.3 Votes allocated in the Council

Member state	Votes
France	10
Germany	10
Italy	10
UK	10
Spain	8
Belgium	5
Greece	5
Netherlands	5
Portugal	5
Sweden	4
Austria	4
Denmark	3
Finland	3
Ireland	3
Luxembourg	2
Total	**87**

Council is required to act by a qualified majority, the votes accorded each member state in Council are as given in Table 2.3.

For their adoption, Acts of the Council require the following:

- at least sixty-two votes in favour where the Treaty requires them to be adopted on a proposal from the EC
- sixty-two votes in favour, cast by at least ten members, in other cases.

It should be noted that abstention by members does not prevent the adoption by the Council of Acts that require unanimity. Over the last decade or so, the areas in which QMV is sufficient for a decision to be taken have been extended gradually. The Amsterdam Treaty looked to enlargement and extended the areas where QMV is sufficient. However, unanimity is still required on constitutional matters and for highly sensitive areas such as taxation.

2.8 The European Parliament

The EP's primary objectives are to pass good laws and to scrutinise and control the use of Executive power (by the EC). The EP is directly elected by the peoples of each member state, elections being held every five years. There are 626 MEPs from the fifteen member countries, the absolute majority therefore being 314 (see Table 2.4).

Originally, the Treaty of Rome only gave the EP a consultative role, allowing the Commission to propose and the Council of Ministers to decide legislation. Subsequent Treaties have extended Parliament's influence to amending and even adopting legislation, so that Parliament and the Council now share the power of decision in a large number of areas. The Parliament's responsibilities and powers

Table 2.4 Population, number of MEPs and Commissioners per member state

1997	Population (× 1,000)	MEPs	Members of the Commission
Germany	81,599	99	2
UK	58,606	87	2
France	58,198	87	2
Italy	57,301	87	2
Spain	39,210	64	2
Netherlands	15,459	31	1
Greece	10,454	25	1
Belgium	10,137	25	1
Portugal	9,917	25	1
Sweden	8,827	22	1
Austria	8,047	21	1
Denmark	5,228	16	1
Finland	5,108	16	1
Ireland	3,598	15	1
Luxembourg	410	6	1
Total	**372,099**	**626**	**20**

were increased by the Single European Act, the Treaty of the European Union and the Amsterdam Treaty, the latter giving the EP a role in deciding the Presidency of the EC. The President of the Commission is nominated by the heads of state and government, but the appointment only becomes effective after it has been endorsed by the EP. The EP also has a role in approving the Commissioners chosen by the President and the national governments. The Amsterdam Treaty recognised that differences between member states will become more marked with enlargement. The Treaty makes 'closer cooperation' possible, while safeguarding the Community's objectives and preventing a situation where slow movers can never catch up with the vanguard.

There are eight main political groups in the EP (see Table 2.5). Each MEP sits on at least one of twenty or so Parliamentary committees, the most important of these in a food context being the Committee on the Environment, Public Health and Consumer Protection. Such committees produce suggested amendments to and opinions on EC proposals, which are referred to one of the monthly plenary sessions of the EP where they are discussed, amended and finally adopted by all MEPs.

2.9 The Court of Justice and the Court of Auditors

The Court of Justice (CoJ) works to ensure that the law is observed in terms of the interpretation and application of the various Treaties and generally in all of the activities of the EU. The Court of Auditors (CoA) is responsible for checking that the EU spends its money according to its budgetary rules and regulations and for the purposes for which it is intended.

Table 2.5 The eight main political groups in the EP

Group	No. members
Group of the European People's Party (EPP), Christian Democratic Group	233
Socialist Group (PSE)	180
Liberal Democratic and Reform Group (ELDR)	50
Greens (V) European Radical Alliance (EFA)	48
United Left (GUE/NGL)	42
Union for Europe Group (UEN)	30
Group of Independents for a Europe of Nations (IND)	27
Europe of Democracies and Diversities Group (EDD)	16

2.10 ECOSOC and the Committee of the Regions

The Economic and Social Committee (ECOSOC) is an advisory body of 222 members drawn from the economic and social interests in Europe. It was established by the 1957 Treaty of Rome to assist in ensuring that the positions of the EU's various economic and social categories are taken into account. Its main role is to issue opinions on draft Community legislation, being referred to it by the EC and the Council. Members are nominated by member states' governments and are appointed by the Council for a renewable four-year term of office, the current term of appointment being to 2002. Members are divided into three groups (employers (Group I), workers (Group II) and Various Interests (Group III)). France, Germany, Italy and the UK each have twenty-four members, Spain has twenty-one, Austria, Belgium, Greece, the Netherlands, Portugal and Sweden each have twelve, Denmark and Finland nine and Luxembourg six. A bureau of thirty-six members (twelve per Group) is elected every two years, and a president and two vice-presidents are chosen from each of the Groups on rotation. ECOSOC has six sections:

1. Agriculture, Rural Development and the Environment.
2. Economic and Monetary Union and Economic and Social Cohesion.
3. Employment, Social Affairs and Citizenship.
4. External Relations.
5. The Single Market, Production and Consumption.
6. Transport Energy, Infrastructure and Information Society.

Consultation of ECOSOC by the EC or Council is mandatory in certain cases; in others it is optional. The Single European Act (SEA) and Maastricht Treaty extended the range of issues on which ECOSOC must be consulted. The Amsterdam Treaty further increased the range of issues and allowed it to be consulted by the EP. On average it delivers 170 advisory documents and opinions a year, of which 15% or so are on its own initiative. As a rule, ECOSOC meets ten times a year. Opinions are adopted at the plenary sessions by simple majority.

All opinions are forwarded to the EU's decision-making bodies and then published in the *Official Journal of the European Communities* (*OJ*).

The Committee of the Regions (CoR) also has 222 members appointed until 2002. They were created to give expression to an 'ever closer Union' of European nations. As the Union's responsibilities have broadened, the institutions have grown larger and more numerous. In its first twenty years, the Commission would propose, the Parliament would advise, the Council of Ministers would decide and the Court of Justice would interpret. In the last twenty years, the Parliament has become directly elected and acquired new powers, the European Court of Auditors has arrived on the scene, the European Investment Bank has emerged as a major source of finance for economic development, the Economic and Social Committee has testified to the value of debate and cooperation between the economic and social partners and, most recently, the Committee of the Regions has been set up to advance regional interests and diversity.

2.11 Legislative process

The legislative process always begins with a proposal from the EC. Before it issues a draft item of legislation, the EC carries out preliminary soundings and discussions with representatives of governments, industry, the trade unions, special interest groups and, where necessary, technical experts. The EC has to send its proposal for legislation formally to the Council of Ministers and the EP. These institutions then work together to produce final legislation (Fig. 2.3).

In agreement with the EC, the Council can amend a proposal by a qualified majority vote, but if the EC does not agree, the change requires unanimity. The EP shares the power of co-decision with the Council in most areas, and has to be consulted in others. When revising its proposals, the EC is required to take the EP's

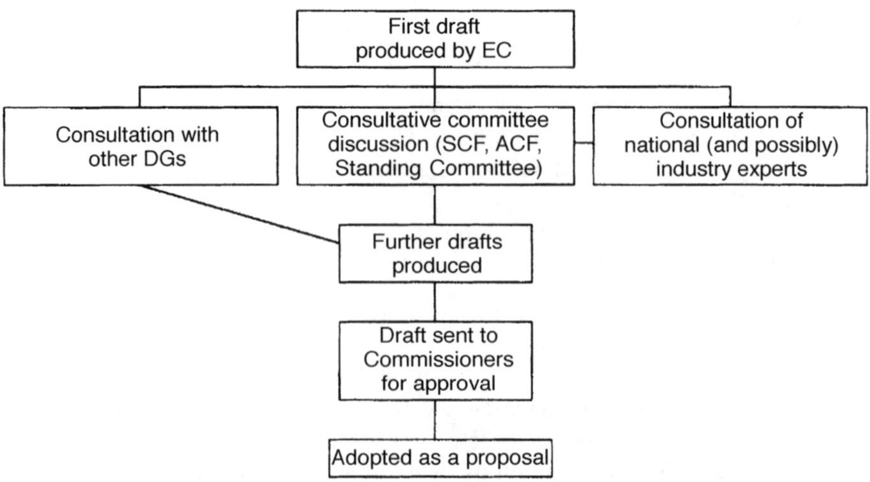

Fig. 2.3 Development of an EC proposal for legislation.

amendments into consideration. Subsidiarity is enshrined in the Treaty on European Union and is applied by the EC in such a way as to ensure that the EU takes action only when it will be more effective than if left to individual member states.

2.11.1 The co-decision procedure (Fig. 2.4)

The Maastricht Treaty gave the EP the power of 'co-decision' with the Council in a limited number of areas such as research, health and culture. The co-decision procedure shares decision-making equally between the EP and the Council. A conciliation committee made up of equal numbers of MEPs and Council representatives, with the Commission present, seeks a compromise on a text that the Council and EP can both subsequently endorse. If there is no agreement, Parliament can reject the proposal outright.

The Amsterdam Treaty increased the EP's responsibilities by making the co-decision procedure the general rule in policy areas apart from Economic and Monetary Union (EMU), applying to a large number of areas including consumer protection and health and most legislation relating to food. The co-decision procedure can be finalised within two years, although double this length of time is normal.

2.11.2 The cooperation and consultation procedures

The Council has the final say on a significant number of other policy areas. The EP can either amend the Council's draft legislation (the 'cooperation' procedure) or withhold its assent to Council decisions in certain areas (e.g. residence rights, Treaties of Accession). The cooperation procedure allows the EP to improve proposed legislation by amendment. It involves two readings in the EP, and currently only applies to EMU issues.

The consultation procedure requires an opinion from the EP before a legislative proposal from the Commission can be adopted by the Council (see Fig. 2.5).

2.12 Forms of legislation: directives, regulations and decisions

There are three main types of European legislation, each fulfilling a specific function. All agreed legislative documents must be published in the *Official Journal of the European Communities*, commonly known as the 'OJ', in order to become law.

2.12.1 Directives

A directive describes compulsory objectives but allows member states flexibility in its translation into national law through national 'implementing' legislation,

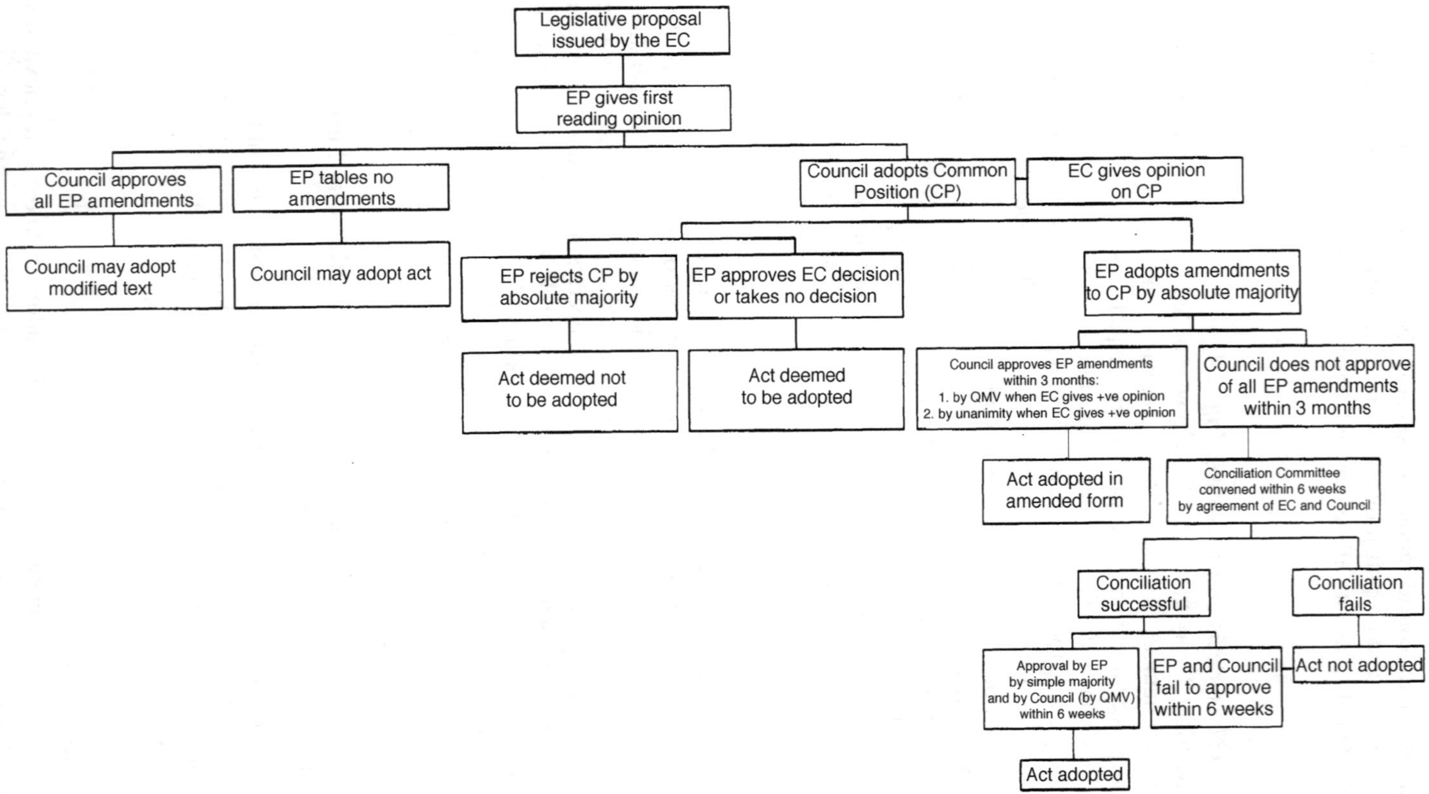

Fig. 2.4 The co-decision procedure.

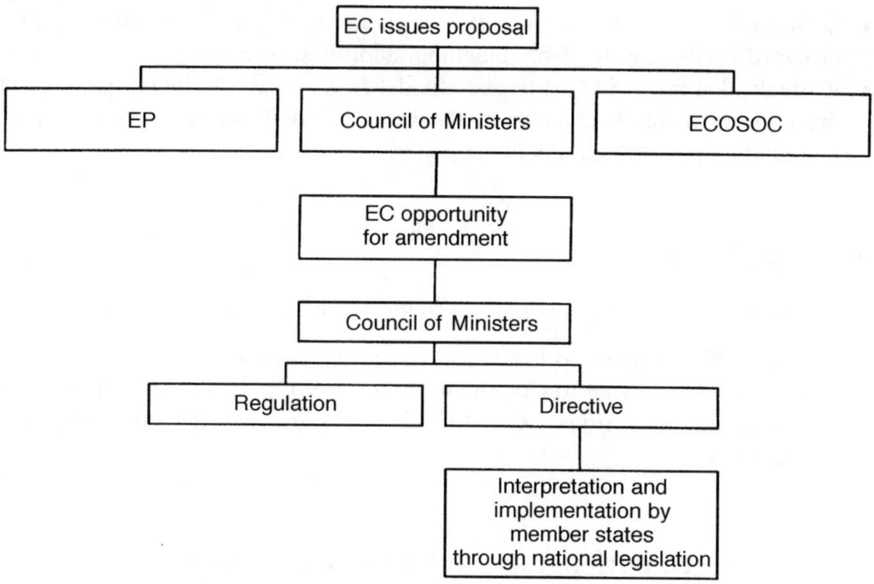

Fig. 2.5 The consultation procedure.

which usually must be carried out within twelve to eighteen months of the directive coming into force. A directive may allow specific derogations (exceptions), enabling a member state to take account of existing national laws or practices, but each directive states when it becomes forbidden to allow non-complying products on the market.

2.12.2 Regulations

A regulation applies directly to all member states and is binding as soon as it is adopted and published in the *OJ*. A regulation is not required to be transposed into national law through national legislation, therefore removing elements of flexibility of directives. Regulations can be issued by the Council, by the Council and the EP acting jointly or by the EC when it is given the power to take action on its own.

2.12.3 Decisions

A decision is binding on the parties to whom it is addressed. These may be member states, companies or individuals.

2.13 Horizontal or vertical legislation?

European legislation is often referred to as being 'horizontal' or 'vertical', meaning:

- horizontal: dealing with a particular aspect of law applicable to all foods or a group of foods (e.g. hygiene, labelling, additives, packaging)
- vertical: dealing with particular foods (e.g. meat products, jams) and prescribing standards controlling all aspects of the food (e.g. raw materials, ingredients, processing, labelling).

2.14 Publication

Once adopted, legislation is published in the *OJ*, of which there are two series:

1. Series – 'L' referring to formal published legislation.
2. Series – 'C' referring to Communications. The C series of the *OJ* contains information on various EC activities and sometimes contains proposed legislation.

2.15 The EU, national and international regulation

A country joining the EU is compelled to implement EU legislation in order to seek to avoid barriers to trade with its EU partners, and to ensure that they are working to equivalent technical standards that will protect consumer health. Courts within the EU member states must use an EC regulation as if it were national law as EC regulations are more powerful and override any national provision with which it may be in conflict. The food and other industries must comply with EC regulations even if there has been no national law relating to it.

In international food law the two most important regulatory bodies are the World Trade Organisation (WTO) and the Codex Alimentarius Commission, often simply referred to as 'Codex'. Codex, which is run under the aegis of the United Nations (UN) Food and Agricultural Organisation (FAO) and the World Health Organisation (WHO), was formed in 1962, to facilitate the development of trade in foodstuffs. It has developed indicative standards, recommendations and guidelines aimed at food safety and fair trade. Its members are national governments and the EC is represented at its meetings on behalf of the EU as a whole. EU legislation relates to the WTO and is, where relevant, in line with Codex requirements.

The key role of Codex in the development of international trade standards was recognised when the WTO was established in January 1995, updating and replacing the General Agreement on Tariffs and Trade. The 'General Agreement' setting up the WTO was supplemented by several more detailed agreements including the Agreement on Sanitary and Phytosanitary Measures (the 'SPS' Agreement) and the Agreement on Technical Barriers to Trade (the 'TBT' Agreement). Codex standards are recognised as the basic standard upon which national food safety measures of SPS member countries should be based, and are therefore particularly relevant where cases of trade dispute are brought to the WTO's disputes panel. SPS members are required by the Agreement to accept the sanitary and phytosanitary measures of other members as being equivalent, even if these measures differ from their own, if the exporting member objectively demonstrates to the importing member that its measures achieve the importing member's appropriate level of sanitary or phytosanitary protection.

2.16 How EU law works: an example

Council Directive 93/43/EEC of 14 June 1993 on the hygiene of foodstuffs, the Food Hygiene Directive (FHD) as it is commonly known, is in the process of being replaced by a consolidation and simplification of all food hygiene legislation including seventeen so-called 'vertical' directives which relate to specific protein product and material areas such as meat of various animal origins, eggs and fishery. The FHD aimed to harmonise general hygiene rules for the preparation, processing, manufacturing, packing, storing, transportation, distribution, handling and offering of food for sale or supply to the consumer. In doing this, the FHD, for the first time, introduced concepts from HACCP (Hazard Analysis and Critical Control Points) and risk assessment into horizontal food law, drawing back from legislation simply prescribing detailed statutory requirements relating to specific practices, although certain core requirements (e.g. general requirement for food premises) were given in the ten appendices to the Directive. The FHD specifically referred in Article 5 to the production of voluntary guides to good hygiene practice for food businesses as a guide to compliance with Article 3 of the Directive, which related to food operations being carried out in a 'hygienic way' and that HACCP principles be used in doing so. Article 5 also referred directly to the CODEX recommended International Code of Practice, General Principles of Food Hygiene, stating that guides shall, where appropriate, have regard to it.

The FHD opened the way for further requirements on microbiological and temperature control criteria for certain classes of foodstuffs; however, these have not yet been manifested in European rules. The FHD has meant that since its latest implementation date of 14 December 1995, food sector businesses have been required to identify 'any step in their activities which is critical to ensuring food safety and ensure that adequate safety procedures are identified, implemented, maintained and reviewed' on the basis of five of the seven principles of HACCP. In the UK, the FHD was implemented through the Food Safety (General Food Hygiene) Regulations 1995 (SI 1763), which came into force on 15 September 1995. The Regulations implemented the FHD both in spirit and detail in the main but, perhaps most notably, added detail regarding guides to good hygienic practice referred to in the Directive, requiring them to be developed in accordance with a template (DH, 1995) developed by the UK Department of Health, which was the lead government department in this legislation. This unique UK development illustrates well one of the problematic aspects of using directives, i.e. they open the way for variation in interpretation and implementation.

Since 1993 and the FHD, the regulatory climate has continued to develop, with emphasis on HACCP and taking a whole-chain approach to food safety controls. A consequence of this and the need to simplify the myriad hygiene directives and regulations is resultant work on the consolidation of food hygiene legislation at all stages of the food chain. Consolidation will be through a series of regulations and not as directives, since the EC wishes to ensure consistency of application throughout the EU. Four linked proposals for regulations on food safety rules and associated animal health control have been issued at the time of writing:

1. Regulation of the European Parliament and of the Council on the hygiene of foodstuffs (EC, 2000/0178 (COD)).
2. Regulation of the European Parliament and of the Council laying down specific hygiene rules for food of animal origin (EC, 2000/0179 (COD)).
3. Regulation of the European Parliament and of the Council laying down detailed rules for the organisation of official controls on products of animal origin intended for human consumption (EC, 2000/0182 (COD)).
4. Council Regulation laying down the animal health rules governing the production, placing on the market and importation of products of animal origin intended for human consumption (EC, 2000/0180 (COD)).

An EP Directive (200/0181 (CNS)) is also proposed to repeal certain existing directives on the hygiene of foodstuffs and the health conditions for the production and the placing on the market of certain products of animal origin. Further, detailed information on hygiene legislation is contained in Chapter 3.

2.17 References and further reading

DH, 1995, 'A template: industry guides to good hygiene practice', Department of Health, London, UK.

EC, undated, 'A brief introduction to the EU', European Commission.

EC, 1996, 'Serving the European Union: a citizen's guide', Luxembourg, Publications Office of the European Communities.

EC, 1997, 'Commission Decision setting up the Scientific Committees in the field of consumer health and food safety', Commission Decision No. 97/579/EC of 23 July 1997, OJ L237 of 28 August 1997.

EC, 1997, 'Inter-services operations manual establishing co-operation procedures between Directorate General III, V, VI and XXIV', European Commission.

EC, 2000, 'Qualified majority voting', Factsheet, European Commission.

EC, 2000, European Commission Proposals to Consolidate and Simplify EU Legislation – various proposals, Commission of the European Communities, COM 2000 438 final.

Part I

Food safety

3

Hygiene

M. Fogden, Meat and Livestock Commission, Milton Keynes

3.1 Introduction

A series of food scares has reduced consumer confidence in food safety even though the risk from food is generally extremely low. It is important to reassure consumers and restore their confidence. This requires elimination of the basis for their concern, by the industry promising and providing safe food with the application of quality management systems that will guarantee this. The industry is achieving this, and independent auditing of these systems to demonstrate their performance is becoming increasingly common.

Appropriate hygiene must be applied as necessary during all stages preceding the consumption of food to ensure that it is safe. It is apparent that this, and improved public awareness of it, are fundamental to the maintenance of consumer confidence. It also aids business profitability by reducing losses. Such efforts will not, however, prevent illness caused by subsequent unhygienic consumer activities.

There must be an adequately equipped and controlled environment and appropriate hygiene procedures for the production, handling, storage, distribution and supply of food ingredients, packaging materials and foods. This may be based on detailed prescriptive controls providing a rigid guarantee of safe working, or a more flexible management system based on the control of objectively assessed risk, or a combination of these. In each case, implementation must be under the control of food business operators, who are responsible for ensuring that the products they supply are safe. A regulatory regime with effective enforcement is also necessary to deal with residual errors, failures and especially abuses.

This chapter expresses the personal views of the author and must not be attributed to MLC.

The nature and application of this regime is the topic of this chapter. It covers the structure of the control system, before examining the EU legal requirements. There is legislation generally applicable to retailing and catering for all foods, and to the whole supply chain for many foods. There are also specific requirements applicable only to the production of foods of animal origin on an industrial scale and in those smaller businesses that are caught by these rules and therefore require similar controls to be in place. The chapter then considers future trends before providing a short list of sources of further information. Other chapters also include hygiene-related information.

3.2 Hygiene regulation in the EU: key themes

From early in the development of the European Community, its member states moved towards harmonised food hygiene control through Community laws. Attention was given initially to the more perishable commodities, particularly when they cross frontiers between those states.

3.2.1 Horizontal and vertical control measures

The European Commission developed legislation for products of animal origin within the Common Agriculture Policy, in a set of 'vertical' directives, each covering a restricted range of foodstuffs, usually in considerable detail and including some non-sanitary matters. They contain numerous inconsistencies, often for no obvious technical reasons (Fogden 1994–96).

The existing Community hygiene controls on products of animal origin were reconsidered during the period around 1990 when the single market was being created. With the elimination of border controls, there was concern that food obtained under less stringent national rules could enter other states without further checks or controls. It was decided to harmonise the national production and trade requirements to a similar standard to eliminate this, so existing directives were updated. A 'horizontal' directive providing general hygiene rules for matters and foods not covered by the vertical legislation was added.

Hygiene rules cannot be considered satisfactory unless those concerned in their application and enforcement can interpret them effectively and consistently. They must be capable of ensuring the protection of public health and should be adequately flexible to satisfy diverse but essential needs. In many cases this is the situation, but improvements are possible.

Thus a group of directives was adopted to ensure hygienic production and marketing of all foods. There were difficulties (e.g. with proposals for minced meat hygiene controls – see section 3.6), but most vertical measures were adopted by September 1992 and the horizontal directive on the hygiene of foodstuffs (93/43/EEC) followed in June 1993. The latter is enforced under national food control systems while the vertical rules are under veterinary control.

Legislation also covers the importation of foodstuffs from third countries into the Community, with a series of decisions listing the individual establishments that have been approved.

3.2.2 The scope of regulation: what is hygiene?

Article 2 of the horizontal 'General Food Hygiene Directive' (93/43/EEC) defines 'food hygiene' as 'all measures necessary to ensure the safety and wholesomeness of foodstuffs' and applies during 'all stages after primary production', this including harvesting, milking and slaughter. Circuitously and somewhat unhelpfully, it then defines 'wholesome food' as that 'which is fit for human consumption as far as hygiene is concerned'.

The vertical hygiene directives are primarily aimed at controlling hygiene but include other rules that target the control of quality and the provision of information to a purchaser through labelling. Such aspects are certainly important in their own right in ensuring good product quality and in providing information and assurance to consumers about the foods that they intend to consume, but they do not always fit within 'hygiene' as defined above. The juxtaposition of these elements can be confusing (Fogden 1994–96, Part 7), especially as they were developed by specialist veterinary officials with a limited understanding of general food law. Some of these initiatives are worthy but if specific controls are needed, they would be better placed outside these hygiene directives. Many are already covered in principle in horizontal directives, for example in the food labelling directive (2000/13/EC), which requires food to be labelled appropriately and in accordance with general and/or detailed rules. A review has addressed these concerns (see sections 3.5.3 and 3.7.1).

3.2.3 Rigid control systems or risk management

Hygiene rules must be applied broadly to the production of food and its supply chain to provide effective protection against food safety problems. Moreover, operators should not confine themselves to compliance with legislated generic hygiene measures but should also consider whether additional precautions or control systems are required in the particular circumstances of their businesses.

Increasingly, risk management systems are being introduced. These are commonly based on the Hazard Analysis and Critical Control Points (HACCP) system developed originally for microbiological control of foods intended to be consumed in American space missions. A comprehensive, properly implemented risk management system based on HACCP can make a very significant contribution to ensuring food safety (see sections 3.2.4 and 3.5.3).

Some hygiene directives demand risk management, to different extents, but many vertical directives rely on rigid requirements specified in considerable detail. These cover all businesses in that category, rather than permitting controls that are adequate and sufficient for particular circumstances. These provide no encouragement to an operator to introduce appropriate risk management systems with different and probably less onerous controls since these must be introduced in addition to the prescribed requirements. Other directives apply a HACCP-based procedure on top of prescriptive controls specified to varying levels of detail and complexity.

Hygiene deals with the preservation of health and a hygienic business should control the risk of illness resulting from the operations carried out on its premises. Implementation of the necessary controls also gives advantages in maintaining product quality. There are three main requirements:

1. Avoid contamination of the food in the first place.
2. Avoid the spread of contamination.
3. Eliminate harmful contamination.

3.2.4 Hazard Analysis and Critical Control Points (HACCP)

HACCP is recommended by leading health authorities including the WHO/FAO Codex Alimentarius Commission ('Codex') as the basis for hygiene risk management. Specialist texts and advice on HACCP are readily available. It is a seven-stage system which examines the production process and determines the critical points that need to be controlled in order to ensure food safety. The seven principles of HACCP are as follows (Codex Alimentarius 1997b):

1. Conduct a hazard analysis.
2. Determine the critical control points (CCPs).
3. Establish critical limit(s).
4. Establish a system to monitor control of the CCPs.
5. Establish the corrective action to be taken when monitoring indicates that a particular CCP is not under control.
6. Establish procedures for verification to confirm that the HACCP system is working effectively.
7. Establish documentation concerning all procedures and records appropriate to these principles and their application.

The Codex HACCP Code gives further guidance. This includes the following:

- A food chain sector should already be operating according to Codex General Principles of Food Hygiene (Codex Alimentarius 1997a), other appropriate Codex Codes of Practice and food safety legislation before the application of HACCP.
- Management commitment to HACCP is essential for the implementation of a HACCP system.
- Redesign of an operation may be necessary if a hazard requiring control is identified but no CCP can be found.
- Each operation should be subject to HACCP, and reviewed as necessary.
- Be flexible in applying HACCP, taking account of all the circumstances.

HACCP is not formally required as such by any EU food hygiene legislation at present, although substantial parts of the principles of HACCP are incorporated in some areas, including the General Food Hygiene Directive (section 3.4.2) and the directives controlling meat preparations and products. In general, however, vertical EU controls are based on prescriptive detail rather than self-control. The attitude and/or knowledge required for effective self-control of hygiene risks is lacking in some food businesses, and it is likely that at least some prescriptive rules will continue to form the basis of legislated requirements for some time.

However, the global trend is towards self-regulation, and it is appropriate to provide a legislative system that permits this for businesses that can demonstrate relevant competence and effectiveness. These could then profit from derogations from the prescriptive requirements, giving them flexibility in the system they introduce and avoiding unnecessary expense occasioned by redundant measures. It is easier to enforce detailed rules than to assess individual systems of control, so inspectorates need to be trained to ensure that they are able to satisfy themselves that food hygiene standards are being met (section 3.7.3). This is already a problem, since there is a requirement in the General Food Hygiene Directive for a HACCP-based system to be in place. Such systems are currently weak at best in many premises where there is an apparent lack of understanding, competence or application. There is still a considerable need for education and encouragement, probably before resorting to strong enforcement (except in dangerous situations). The so-called 'honeymoon period' cannot, however, go on for ever.

3.3 Enforcement of hygiene regulations

The nature of EU directives is that they have to be implemented through national legislation, unlike its regulations and decisions which apply automatically. Each member state must introduce its own measures to implement each directive within a specified period, to achieve the objectives agreed and set out in the directive. So, for example, in Britain the General Food Hygiene Directive has been implemented by the Food Safety (General Food Hygiene) Regulations 1995 which largely repeat the directive's provisions but are drafted according to the national legal tradition.

The vertical directives were originally proposed as regulations. However, the member states decided not to control food hygiene in this inflexible manner but as directives, the form proposed for the horizontal measure. These allow governments to implement the controls, meeting the objectives, in ways that suit national or cultural preferences.

There are opportunities for inconsistencies. Harmonised rules can be introduced effectively through directives but the result is less uniform than when regulations are introduced directly and simultaneously into each state.

This, with possible variability of enforcement, can result in unfair competition and protectionism. The Commission monitors the position to avoid this.

3.3.1 Official control

National governments are required by the Official Control Directive (89/397/EEC) to enforce food hygiene legislation. This is devolved in many states to a local level through municipal or regional authorities, indicating that a coordinating system should be in place to improve consistency. National enforcement officials interact on a European basis through the Food Law Enforcement Practitioners' forum.

Official inspections of production and supply establishments are often supplemented by audits by customers or specialist inspection bodies. These may apply stricter standards than are required by law, and the consequences of failure may be painful and immediate, through loss of business rather than an extended enforcement procedure. In such cases, the official control system can be almost redundant.

3.3.2 Veterinary and non-veterinary enforcement

As indicated previously, the vertical directives are based on veterinary supervision whereas the horizontal directives are not. This can cause difficulties, even friction, where the two systems are controlled by separate national or local authorities. Improved cooperation and administrative coordination would help in some states and it is desirable that legislators improve the interface by reducing some differences between the requirements, which can be confusing.

3.3.3 Civil liability for hygiene failure

It is the responsibility of every business proprietor active in the food chain to ensure that they provide products that will help to ensure consumer safety, whether these be equipment, ingredients or final products.

It is worth remembering that the Product Liability Directive (85/374/EEC) places strict liability on suppliers of all goods, previously excluding primary agricultural products in most member states but now extended to these (Directive 1999/34/EC). This legislation assists injured persons to make a claim for damages in civil law. Claimants need only prove damage to themselves (or their property, subject to a minimum value), and that the damage was caused by a defective product for which the producer was responsible. They do not have to show that there was any fault in what the producer did, or did not do, or that there was any negligence on the part of the producer. Moreover, that liability begins at the end of the chain, with the business supplying the injured person, and passes back down the chain to the original producer of the goods only if each link is able to state from whom they obtained the defective product. Effective traceability is therefore essential for every item purchased by a food business, to transfer liability to the person who is properly responsible for the defect. Appropriate insurance may also be helpful, in case such transfer is impossible.

3.3.4 Consumer aspects

Customer expectations must be met, if food businesses are to thrive. Since hygiene is one of their fundamental demands, satisfying this is clearly high in the list of priorities for business success.

It is believed that a consumer seeks both safe food and confidence that this is being provided. Given that confidence, most will not ordinarily concern themselves with production hygiene. They may well, on the other hand, rightly react strongly against visibly poor hygiene where food is supplied to them. That does not provide any excuse for poor hygiene where consumers are not able to see what is going on; nor is it acceptable to apply hygiene in such places only when an official control inspector is performing an inspection. Expert inspectors are anyway usually able to perceive this.

Regrettably, nowadays consumers in some states are less aware of and less competent in hygiene than previous generations because they have not received sufficient relevant instruction at home or education at school. It is therefore essential that sufficient hygiene instructions are presented in the labelling of food, although manufacturers of food products often properly argue that it is not their task to compensate for lack of general instruction. It is then the responsibility of the consumer to read that information – but regrettably many fail to do so. Better consumer education is necessary in hygiene and in the need to recognise their responsibility in maintaining the hygiene put into the foods and food ingredients they purchase.

Criminal legislation requiring consumers to prepare and serve food hygienically would generally be impractical to enforce and undesirable, except perhaps where gross faults cause serious illness or death (although civil remedies do already exist). It would most probably not reduce significantly the enormous amount of minor food poisoning caused every year by consumers, resulting in discomfort, pain and inability to work.

3.4 The General Food Hygiene Directive (93/43/EEC)

This directive follows the vertical directives' format in comprising a number of articles providing general requirements together with annexed detailed provisions on particular control areas (see Table 3.1).

3.4.1 Essential requirements

These are set out in article 3 of the directive. Fundamentally, article 3(1) requires 'preparation, processing, manufacturing, packaging, storing, trans-

Table 3.1 Structure of the General Food Hygiene Directive

Article	Principal areas covered
1	Scope
2	Defines 'food hygiene', 'food business' and 'wholesome food'
3	Requires hygiene and risk management throughout the food chain
4	Allows for microbiological and temperature control criteria
5	Industry hygiene guides
6	EN 29000 standards
7	Additional national hygiene requirements
8	Official control
9	Enforcement
10	Third country imports – safeguard measures
11	National prevention of health risks
12	Competent authorities
13	Adoption of international standards
14	Adoption of additional requirements
15	Review of implementation
16	Entry into force
17	Applicability to all member states
Annex – detailed requirements	
I–III	Various types of premises
IV	Transport
V	Equipment
VI	Food waste
VII	Water supply
VIII	Personal hygiene in food handling areas
IX	Raw materials, intermediates and finished foods
X	Training

portation, distribution, handling and offering for sale or supply of foodstuffs [to] be carried out in a hygienic way'. In practice, the General Food Hygiene Directive controls most retailers, caterers, the production and supply of all foods that are not of animal origin, and all other food businesses that are not controlled under the vertical directives. Put simply, every part of the food supply chain must be hygienic. Article 3(2) deals with risk management (see section 3.4.2) and article 3(3) requires food businesses to meet the specific hygiene rules laid down in the annex to the directive (see section 3.4.3).

Article 5 allows for the development by the industry of guides to good hygiene practice, in collaboration with interested parties. Codex principles may be applied. If a national competent authority believes that such a guide meets the requirements of article 3, it must forward it to the European Commission which will make it available to the other member states. These guides do not have the force of law for there is quite properly no requirement to follow such guidance, since any other means of complying with the legislation is just as acceptable. However, they have strong persuasive value and proof of compliance with a recognised guide would be very helpful against an enforcement challenge.

European guides may be developed in due course, which would take account of existing national guides but apply throughout the EU.

Article 7 allows member states to introduce national hygiene legislation that exceeds the requirements in the directive provided they do not restrict, hinder or bar intra-Union trade in relevant food. This has been done in some states, which have, for example, introduced temperature controls.

National temperature/time hygiene legislation has not yet been harmonised by the Community. The horizontal directive provides for this in article 4. Currently it specifically requires such control only for microbiologically perishable foods and hot-processed foods which are to be held or served chilled (annex, IX).

3.4.2 Risk management

Article 3(2) requires all food business operators controlled under this legislation to carry out a risk analysis based on the following HACCP principles (it omits the verification and documentation stages):

- analysis of the potential food hazards in a food business operation
- identification of the operational points where food hazards may occur
- decision as to which of the points are critical to food safety (thus establishing the critical control points – CCPs)
- identification and implementation of effective control and monitoring procedures at those CCPs
- review of the analysis and risk management system periodically and when the food business operations change.

This has been a legal requirement since 1996. The law does not currently specifically require documentation of the risk assessment and management system, or of the results of reviews. However, it is suggested that this is in fact a requirement wherever a food business could not reasonably maintain an effective system in place in the absence of documentation. This is thought to be the case in all businesses of any substantial size or operational complexity. Moreover, the presentation of a documented record to an enforcement authority or court is likely to be more persuasive that the requirement had been complied with than unsupported statements. Food businesses are therefore urged to make a reasonable effort to record their assessments, systems and reviews; such documentation may be required in future.

3.4.3 Annex

This provides limited specific controls concerning the structure and facilities in food premises (chapters I–III). Chapter I provides rules applicable to food premises other than movable and temporary ones, which are subject to the rules in chapter III. Chapter II specifies rules that apply to all rooms where food is prepared, treated or processed except dining areas and rooms covered by chapter

III. These chapters require appropriate design and construction of premises to permit good hygiene practices, with temperature control (if necessary), sufficient wash basins and lavatories, ventilation, lighting, drainage and changing facilities. There must be protection against risks of contamination and cross-contamination. The premises, including working surfaces and equipment, must be kept in a sound condition and be easy to clean and disinfect. Additional requirements may result from the application of articles 3(1), 3(2) or 7 (see sections 3.4.1 and 3.4.2).

Chapter IV deals with transport, requiring vehicles and containers to be suitable, with temperature control where appropriate, and in sound condition. They must be able to be cleaned where necessary, especially between loads of different foods, or between foods and non-foods. Mixed loads must be properly separated to avoid contamination. Dedicated receptacles, containers or tankers marked 'for foodstuffs only' must be used to transport food in bulk. Chapter V similarly requires articles, fittings and equipment that come into contact with food to be kept clean, properly maintained and in good condition.

Chapter VI prohibits the unavoidable accumulation of waste in food rooms, and requires food waste and other refuse to be stored in closed or approved containers, again clean, sound, easy to clean and disinfect. Chapter VII requires potable water to be supplied; this must be used wherever necessary to ensure food hygiene, including in the preparation of ice. Other water may be present in the premises (e.g. for steam generation and fire control) but must be kept separate from the potable supply and clearly identified as non-potable.

Personal hygiene is essential. All the other controls will not ensure food safety if the staff contaminate the food because they are dirty, do not wear protective clothing or are liable to transmit diseases. Chapter VIII deals with this, with the second paragraph banning specified people from being allowed to work in food handling areas in a way that could lead to direct or indirect contamination of food with pathogenic micro-organisms. These are people who are known or suspected to suffer from, or be a carrier of, a disease likely to be transmitted through food and also people with infected wounds, skin infections, sores or diarrhoea. It is essential that food business proprietors persuade their staff to declare such incapacities so that they can be put onto tasks where there is no risk of contaminating food. It is important to note that the rule applies to anybody, not just those employed as food handlers, whose presence working in any food handling area puts the food at risk.

Chapter IX prohibits acceptance of raw materials or ingredients that are, or are likely to be, 'so contaminated with parasites, pathogenic micro-organisms or toxic, decomposed or foreign substances' that they would still be unfit for human consumption after passing through normal hygienic sorting, preparatory and processing procedures. Raw materials and ingredients that do enter the premises must be properly stored, handled and used to prevent harmful deterioration and contamination. Food must be protected from contamination that is likely to make it unfit for human consumption, injurious to health or contaminated in such a way that it would be unreasonable to expect it to be consumed in that state. Pests must be controlled. Temperature controls must be in place where this is necessary to prevent a risk to health from the growth of pathogenic micro-organisms or the formation of toxins, although brief periods outside such control are permitted for practical reasons. As mentioned previously, hot foods to be held or served chilled must be cooled as quickly as possible to a safe temperature. Hazardous and inedible substances, including animal feedstuffs, must be adequately labelled and separated in secure containers. In essence, this chapter requires all reasonable precautions to be taken to prevent food being put at risk by contamination during its preparation, storage and handling.

Chapter X is as important as the other provisions. It requires that 'food handlers are supervised and instructed and/or trained in food hygiene matters commensurate with their work activity'. There is no point in having a set of safety rules in place if those working in the premises do not know what is expected of them, and adequate supervision is essential to trap potential failures and other problems. If appropriately carried out, training can instil a sense of ownership into the workforce, and this can be very effective in ensuring that a correct attitude and approach is maintained at all times.

3.5 Specific (vertical) hygiene directives: applicable to particular foodstuffs

These directives apply to the industrial scale production of foods of animal origin but also to some smaller businesses, such as butchers who prepare meat products and sell them to other retailers for onward sale. This can cause problems because of the inflexibility of the requirements.

3.5.1 The legislation

Most EU hygiene and other legislation can be accessed on the Internet (through http://europa.eu.int/eur-lex/). A selection of the principal hygiene directives is listed in Table 3.2 (excluding specialised measures such as those on veterinary residues and transmissible spongiform encephalopathies).

These directives are supplemented by decisions such as those relating to cooked crustacea and molluscs (93/51/EEC) and eggs (94/371/EC).

3.5.2 Areas of control: an example

In general terms, the vertical directives apply to the food chain up to primary processing, which includes harvesting, milking and slaughter. They apply to the industrial production, processing, treatment, inspection, marking, labelling, storage, supply, transportation and related operations, i.e. to the production and 'placing on the market' of various foods of animal origin, but not to their retail sale nor to their supply to consumers by way of catering.

How do the vertical controls apply to the production and placing on the market for human consumption of products of animal origin? The Fresh Meat

Table 3.2 Vertical hygiene directives

Product	Directive	Adopted
Fresh red meat	64/433/EEC	26.6.1964
Fresh poultry meat	71/118/EEC	15.2.1971
Meat products	77/99/EEC	21.12.1976
Egg products	89/437/EEC	20.6.1989
Aquaculture animals/products	91/67/EEC	28.1.1991
Live bivalve molluscs	91/492/EEC	15.7.1991
Fishery products	91/493/EEC	22.7.1991
Rabbit meat and farmed game meat	91/495/EEC	27.11.1990
Wild game meat	92/45/EEC	16.6.1992
Milk and milk products	92/46/EEC	16.6.1992
Fishery products on vessels	92/48/EEC	16.6.1992
Other products of animal origin	92/118/EEC	17.12.1992
Minced meat and meat preparations	94/65/EC	14.12.1994
Animal waste	90/667/EEC	27.11.1990

Directive, 64/433/EEC provides an example. This directive was adopted in 1964 but its text was updated and replaced from 1993 (Directive 91/497/EEC). Temporary derogations were available in the discretion of national authorities (Directive 91/498/EEC) for premises that were unable to comply with the new requirements. Those derogations applied only to structural aspects, not hygiene, and the meat from such establishments had to be distinguished from meat from fully compliant premises.

This directive applies only to the supply of meat from domestic bovine animals, swine, sheep, goats and solipeds. 'Meat' here means all parts of such animals that are suitable for human consumption. 'Fresh meat' means any 'meat' that has not been treated; applying cold treatment to preserve meat, whether or not it is wrapped under vacuum or under a controlled atmosphere, does not count as treating it for these purposes.

It is believed that many of the requirements of this and similar directives could be replaced by a risk analysis and management procedure supplemented by veterinary recommendations.

Premises

The Fresh Meat Directive applies throughout the supply chain, from lairage pre-slaughter, veterinary inspection of the animals through the various stages of production (including cutting, packaging and health marking), to the storage and transportation of the product. It applies to slaughterhouses (abattoirs), cutting plants and cold stores but not to the cutting and storage of fresh meat 'performed in retail shops or in premises adjacent to sale points, where the cutting and storage are performed solely for the purpose of supplying the consumer directly on the spot'. This effectively eliminates most independent butchers' shops from the controls unless they also sell meat to anyone except domestic purchasers and caterers.

A feature of the vertical directives is that they require the national competent authority to approve the premises, equipment and the activities carried out there before the product can be supplied for human consumption. The General Food Hygiene Directive has no such requirement for prior approval.

Structural requirements for establishments producing fresh meat are contained in annex I. Derogation is permitted by article 4 for some establishments based on their limited throughput. This is measured in 'livestock units' (LU) for slaughterhouses (adult bovines and solipeds = 1 LU; other bovines = 0.5 LU; pigs over 100 kg liveweight = 0.2 LU; other pigs = 0.15 LU; sheep and goats = 0.1 LU; lambs, kids and piglets below 15 kg liveweight = 0.05 LU). Slaughterhouses are generally categorised as 'low throughput' if they handle less than 20 LU/week and less than 1,000 LU/year, as are cutting plants producing not more than 3 tonnes of meat per week.

Annex I, chapter I, provides detailed structural requirements covering the quality, cleanliness and condition of walls, floors, drains, changing rooms with lavatories and wash basins, doors, ceilings, insulation, refrigeration, ventilation and lighting, water, hand cleansing and disinfection; taps must not be hand-operable. Tool cleansing and disinfection facilities must be provided, in convenient positions and supplied with water at not less than 82ºC. Protection of meat during loading and unloading is necessary, as are pest control and secure containers or a lockable room to store meat not intended for human consumption – and a lockable room for the exclusive use of the supervising veterinary service. Chapters II, III and IV of annex I provide further requirements for slaughterhouses, cutting plants and cold stores respectively.

Raw materials

The controls on the raw material for the production of fresh meat, animals intended for slaughter, in an approved slaughterhouse are extensive. They are found in articles 3, 4, 5, 6 and 8 and annex I.

Article 3(1) controls the production of carcasses, half carcasses, quarters and smaller cuts, including offal. These must have been obtained from an animal that has satisfied both ante-mortem and post-mortem inspection and is thus shown to be fit for human consumption, while article 5 lists fifteen categories that must be declared unfit for human consumption by the official veterinarian. Article 7 requires meat unfit for human consumption to be clearly distinguished from meat fit for human consumption and to be treated according to the Animal Waste Directive, 90/667/EEC.

Article 6 provides various special controls. Article 8 provides additional controls on veterinary residues. Article 9 requires the presence of a veterinarian in slaughterhouses and cutting plants. Annex I provides detailed specific requirements to ensure the hygiene of raw materials, including structural and storage provisions, and chapter VI deals with ante-mortem inspections.

Operations

Animals must be slaughtered hygienically and under veterinary supervision immediately they are brought into slaughter premises, and thereafter a raft of detailed measures come into effect, intended to guarantee that the meat is fit to eat and protected from contamination. Cutting must take place in an atmosphere that has a temperature not exceeding 12°C; during cutting, boning, wrapping and packaging, the temperature of meat must ordinarily not exceed 7°C.

Carcasses fit for human consumption must be stamped in ink or branded with a health mark under veterinary control in a prescribed manner. Cut meat and offal must be treated similarly, although the mark may be applied to its packaging in certain cases. Only specified colours can be used for health marking.

Products

There are no compositional or labelling controls exceeding hygiene requirements in this directive.

Temperature control

Chapter XIV requires meat to be chilled immediately after post-mortem inspection and kept at a constant internal temperature not exceeding 7°C for carcasses and cuts and 3°C for offal during storage and transportation. Derogations are available from the competent authority for transportation to cutting plants or butchers' shops in the immediate vicinity of the slaughterhouse, provided the meat reaches these within an hour.

If meat is to be frozen, this must be done in the slaughterhouse or cutting plant, or in a cold store to which it was transported directly. It must be cooled without delay to below −12°C and stored below that temperature.

Storage and transport

Conditions are laid down to ensure hygienic storage and transportation. Cut meat and offal must ordinarily be wrapped and packaged unless the wrapping provides sufficient protection, unless it is to be suspended throughout its transport. The veterinarian must ensure that conditions are hygienic, with protected loading and unloading and transport by clean, closed vehicles or containers.

Staff

Annex I requires sufficient changing rooms, with showers, lavatories and wash basins with taps not operable by hands or arms. There must be suitable facilities to wash and disinfect hands near work stations; their taps must not be operable by hand and there must be some hygienic means for drying the hands. It requires 'absolute cleanliness' of staff, and people likely to contaminate meat are prohibited from working on it or handling it. Those working where exposed or wrapped meat is being handled, packaged or transported must wear clean headgear, footwear and working clothes and, where necessary, neck shields or other protective clothing. They must wear clean clothes at the beginning of each working day, renewed during the day as necessary. They must wash and disinfect their hands at each resumption of work and several times during the day. Smoking is prohibited where meat is worked on, handled, stored or transported.

Article 10(3) requires a hygiene training programme to be in place, involving the official veterinarian. Annex III lays down professional qualifications required by auxiliaries assisting the veterinarian.

Management and supervision

The management of the food business is responsible for all aspects of the hygienic operation of the premises. The competent authority, through official veterinarians, is responsible for supervising the operation of the premises and ensuring that it operates hygienically.

3.5.3 Review

A consultation exercise on the consolidation and simplification of the vertical directives was instituted by the European Commission in April 1996, with a second stage in February 1997 which included a draft directive to replace the existing legislation. It has circulated proposals for four regulations and a directive covering the hygiene of foodstuffs and certain animal health rules. It is currently (April 2001) unclear how quickly these may proceed towards adoption although the process is underway as a matter of some priority.

The proposals would base the revised EU hygiene legislation on the General Food Hygiene Directive model, supplemented by specific provisions in areas where additional or more detailed controls are deemed necessary. Specific requirements for the documentation and verification of risk management systems, again based on principles of HACCP, would strengthen this area (see section 3.7.1). They would effect a significant measure of consolidation and simplification although some observers are likely to remain dissatisfied with the extent of this.

3.6 Case study: controversy over minced meat (and meat preparations)

It is essential that meat and foods containing meat are supplied hygienically, to ensure public safety. Regional populations consume minced beef ('mince') in different ways; some invariably cook it thoroughly but elsewhere a significant proportion is consumed lightly cooked or raw. The use of pork and lamb varies. It was not surprising therefore that specific national hygiene requirements varied. Some member states had little legislation; others were restrictive, some limiting mincing to 'on the spot' following a purchaser's request.

Stringent French requirements, developed to restore consumer confidence after hygiene scandals in the 1960s, had formed the basis of a directive applicable in inter-state trade, in 1988. Proposals later that year to extend this to domestic markets were controversial.

Four main issues were isolated that were not directly linked to hygiene and were irrelevant for mince that was to be thoroughly cooked (Fogden 1991). These would increase product costs, affect product quality and cause manufacturing burdens. They involved the following requirements:

- mince must be prepared from meat less than six days old, preventing use of trimmings from matured beef, thus increasing prices and restricting practical production periods
- mince must be chilled within an hour to 2°C, requiring investment in equipment, risking damage to surface tissues and causing significant problems in its transportation – hygiene does not require such stringency.
- only certain parts of the carcass, excluding shin meat (a traditional source) could be minced – this would have increased prices and affected nutritional quality
- frozen meat must be excluded from the production of chilled mince, which made temperature control more difficult and eliminated a traditional practice, causing significant supply problems and increasing prices.

These issues caused special concern in some member states, including the UK and Netherlands, but the measures were demanded by others, particularly France. British estimates suggested an increase of 25% in the price of its mince, primarily affecting vulnerable groups in society. This led to a strenuous debate, eventually resolved in 1994 by permitting certain national derogations from the requirements of a replacement directive (94/65/EC) provided that product safety was not compromised.

3.7 Future trends

The EU hygiene directives are in the process of being reviewed, which should lead to improved consistency and controls.

3.7.1 Review of the directives

There were early calls for the hygiene legislation to be reviewed to eliminate inconsistencies and the European Parliament called for the vertical controls to be subordinated to the horizontal directive. The latter recognised this, requiring the Commission to examine the relationship and, if necessary, make proposals by June 1996. It also had to report and make any appropriate proposals before 1999 on the experience gleaned from the implementation of the horizontal directive.

It seems probable that the horizontal text will eventually be developed to achieve a closer relationship to the vertical directives, as improved after their review (see section 3.5.3), leading to a more consistent package of hygiene legislation. That will take time. The substitution of risk management techniques into the vertical legislation, which is currently based on prescriptive controls, is unlikely to take place soon. The legislators have to satisfy sometimes irrational public demands as well. There are great hurdles to be overcome, but eventually a more consistent and scientific approach is thought to be inevitable.

It is unclear what compromise will be found. However, the current proposals envisage a number of common requirements based on the provisions in Directive 93/43/EEC and including HACCP principles, with specific controls in annexes where these are deemed necessary (whether on hygiene or political grounds). It is hoped that the quality and composition requirements of the vertical directives would be revoked, or transferred to more suitable legislation.

It is thought unlikely that the hygiene directives will be revised into consistent texts and implemented into national legislation before the beginning of 2004. This may be affected by the negotiations intended to lead to the enlargement of the EU.

3.7.2 Discussion

As always, it is essential to refer to the legislated texts to know what is required of any business operator in a particular situation. In the case of EU hygiene legislation, this is more difficult because it is necessary not only to look at the law as enacted in the member state where an operation is taking place, but also to consider the objectives as laid down in the original EU directive which were agreed to by the relevant government as part of that measure when it was adopted. The two usually agree, more or less. But that uncertainty, magnified where more than one state implementation is involved because of trans-frontier activity, can cause problems.

This could have been alleviated by the adoption of regulations having immediate effect rather than directives – but this was politically unacceptable because the governments valued flexibility of approach. The current proposals are for regulations, but these may be adopted as directives, as happened to the proposals that led to the current legislation. Various problems therefore remain, and some issues could have been resolved better.

Improvement of scientific basis

The detailed requirements of the vertical provisions (see section 3.5) contrast with the essential rules in directive 93/43/EEC (see section 3.4), which require adherence to the principles of good hygiene management, although some particular requirements are also specified.

It is not obvious that the use in the vertical texts of the risk management approach of the General Food Hygiene Directive would result in risk to health. That would have eased the task of updating plants and provided flexibility to new establishments. The detail in the vertical directives perhaps suggests a lack of faith by the EU authorities in animal product industry operators and/or in those charged with the official control of this sector. There are grounds for prudence where any perishable foods are being produced, distributed and supplied, but it is strange that such caution is shown during the initial stages of the chain of supply of products of animal origin, whereas later stages, for example in retail outlets, are controlled less repressively, as are other microbiologically sensitive foods.

Consistency, necessity and proportionality

It is essential that hygiene controls are practicable. The current ones are inconsistent, leading to confusion and sometimes to difficulties where more than one applies in an establishment. This requires attention, preferably resulting in technically justifiable controls.

These controls must satisfy their purpose, to control hygiene so that public safety can be assured. This is generally the case, but sometimes excessive requirements have been introduced, breaching the EU principle of proportionality that should apply to prevent legislation in excess of what is required to solve a problem. These often just satisfy political needs by enabling governments to avoid reducing unjustifiable controls, because of their fear of alarming consumers. In an EU context, politics often outweighs science when legislative compromises have to be agreed.

Elimination of other measures

It is suggested that there is no place in hygiene legislation for non-hygiene controls. There are many that should be moved elsewhere, or preferably eliminated in some cases, to ensure proper attention to others that do ensure food safety. Moreover, these often seem to have been introduced in the vertical texts without understanding that existing horizontal controls, for example in food labelling directives, are adequate. In this context, it may be noted that the current proposals define 'food hygiene' as 'all measures and conditions necessary to ensure food safety and fitness for human consumption of foodstuffs according to their probable and/or intended use'.

3.7.3 Outcome: self-regulation or prescription

Risk analysis and management provide a mechanism that can ensure food safety equally as well as prescriptive legislation. Both approaches require commitment and/or enforcement to be effectively implemented. What is required is a positive, competent and thorough attitude and implementation of good hygiene standards by everyone involved in every business in the food supply chain. This cannot be instituted by legislation, nor is it likely. It can be improved by educating people into understanding why it is essential, and what consequences can follow failure.

Confidence in industry management

At present, there is reason to lack faith in some food businesses. Their hygiene control is inadequate, putting consumer safety at risk. Consequently there is, and there will remain, a need for prescriptive legislation supported by effective enforcement and penalties.

Many businesses, however, are being run well. There is scope for these to benefit from relaxation of prescriptive detail, allowing them to improve their performance and profitability in a more flexible manner. It is suggested that this should only be done where the enforcement authority is satisfied that the attitude and technical competence in the business is such that it will maintain high hygiene standards. It should be possible for all businesses to benefit from this, in principle, and the authorities would clearly need to maintain an adequate level of surveillance to ensure that the situation remains acceptable.

Ease of enforcement

However, failure to implement risk management systems is more difficult to enforce than failure to comply with detailed requirements. There is a need to employ thorough and thoroughly competent officials with a good understanding of hygiene as it applies in the particular businesses that they inspect. Even then, problems arise because hygiene practices are often debatable and faults can be difficult to challenge objectively so as to satisfy a court. It is therefore probably wise to err slightly on the side of caution in the public interest for all businesses handling any perishable foods, not just those that handle such products of animal origin. However, those able to demonstrate a history of good attitude and control should be permitted to manage their hygiene in a business-efficient way.

Table 3.3 European trade associations

Sector	Organisation abbreviation	Telephone
Agriculture	COPA and COGECA	+32 (0)2 287 27 11
Aquaculture	EAS	+32 (0)59 32 38 59
Bakery and confectionery	CEBP	+32 (0)2 230 34 16
Butchers	IBC	+32 (0)2 230 38 76
Butter	TRANSBEUROP	+32 (0)2 230 44 48
Milk and dairy products	EUCOLAIT	+32 (0)2 230 44 48
Eggs, game and poultry	EUWEP	+31 (0)30 69 67203
Fish	EUROPECHE	+32 (0)2 230 48 48
Fish processors	AIPCEE	+32 (0)2 743 87 30
Food and drink	CIAA	+32 (0)2 514 11 11
Fruit and vegetable nectars	AIJN	+32 (0)2 743 87 30
Ice cream	EUROGLACES	+33 (0)1 53 42 13 38
Livestock and meat	UECBV	+32 (0)2 230 46 03
Mayonnaise and sauces	CIMSCEE	+32 (0)2 743 87 30
Meat processors	CLITRAVI	+32 (0)2 203 51 41
Poultry and game	CDVGP	+32 (0)2 512 61 78
Poultry and poultry processing	AVEC	+45 (0)33 25 41 00
Processed cheese	ASSIFONTE	+49 (0)228 95 96 90
Retailing	EUROCOMMERCE	+32 (0)2 230 58 74
	FEMGED	+32 (0)2 734 32 89
	GECODE	+49 (0)221 936 55770
Soft drinks	UNESDA	+32 (0)2 743 40 50
Tomato products	OEICT	+32 (0)2 743 87 30

3.8 Sources of further information and advice

3.8.1 Trade associations

The selected organisations in Table 3.3 perform representative functions for national associations and individual companies at European level. See also the *European Public Affairs Directory*.

3.8.2 Consumer groups

European Bureau of Consumers' Unions (BEUC) Tel. +32 (0)2 735 31 10
European Federation of Consumers' Co-operatives (Eurocoop) Tel. +32 (0)2 230 32 44

3.8.3 Other contact points

European Commission, Rue de la Loi 200, B-1049 Brussels, Belgium Tel. +32 (0)2 299 11 11.
Food Law Enforcement Practitioners' forum Tel. +31 (0)70 340 50 60
Meat and Livestock Commission Tel. +44 (0)1908 677577 or +32 (0)2 230 86 68
National Agriculture, Consumer Protection and Health authorities

3.9 References

3.9.1 EU legislation (in the Official Journal)

Number	OJ	Date	Page
Directives			
64/433/EEC	121	29 July 1964	2,012
71/118/EEC	L 55	8 March 1971	23
77/99/EEC	L 26	31 January 1977	85
80/778/EEC	L 229	30 August 1980	11
85/374/EEC	L 210	7 August 1985	29
88/657/EEC	L 382	14 December 1988	3
89/397/EEC	L 186	30 June 1989	23
90/667/EEC	L 363	27 December 1990	51
91/67/EEC	L 46	19 February 1991	1
91/492/EEC	L 268	24 September 1991	1
91/493/EEC	L 268	24 September 1991	15
91/495/EEC	L 268	24 September 1991	41
91/497/EEC	L 268	24 September 1991	69
91/498/EEC	L 268	24 September 1991	105
92/5/EEC	L 57	2 March 1992	1
92/45/EEC	L 268	14 September 1992	35
92/46/EEC	L 268	14 September 1992	1
92/48/EEC	L 187	7 July 1992	41
92/118/EEC	L 62	15 March 1993	49

93/43/EEC	L 175	19 July 1993	1
94/65/EC	L 368	31 December 1994	10
98/83/EC	L 330	5 December 1998	32
1999/34/EC	L141	4 June 1999	20
2000/13/EC	L109	6 May 2000	29
Decisions			
93/51/EEC	L 13	21 January 1993	11
94/371/EC	L 168	2 July 1994	34

3.9.2 Other publications

CODEX ALIMENTARIUS (1997a) *Recommended International Code of Practice: General Principles of Food Hygiene*, FAO/WHO, Rome.

CODEX ALIMENTARIUS (1997b) *Hazard Analysis and Critical Control Point (HACCP) System and Guidelines for its Application*, FAO/WHO, Rome.

European Public Affairs Directory, Landmarks SA, Brussels.

FOGDEN, M. (1991) 'European Community minced meat legislation', *European Food Law Review*, 2–3, 150, Frankfurt am Main, Germany

FOGDEN, M. (1994–96) 'European Community food hygiene legislation', Part 1 'Introduction and preambles – the principles', 1994 (5) 65; Part 2 'Areas of control', 1994 (5) 181; Part 3 'Structural requirements', 1994 (5) 267; Part 4 'Controls on raw materials', 1994 (5) 367; Part 5 'Processing controls', 1995 (6) 21; Part 6 'Controls on finished products', 1995 (6) 32; Part 7 'Composition, labelling and microbiology', 1995 (6) 379; Part 8 'Administration and personnel', 1996 (7) 15, *European Food Law Review*, Frankfurt am Main, Germany.

LUGT, M. (1999) 'Enforcing European and national food law in the Netherlands and England', Koninklijke Vermande BV, Netherlands.

SPRENGER, R. (1991) *Hygiene for Management*, Highfield Publications, Doncaster, UK.

4

Additives

D. W. Flowerdew, Consultant (formerly Leatherhead Food RA)

4.1 Introduction

A directive that sets safety and need criteria for the use of additives in food and makes provisions for further specific rules to be drawn up provides the basic general controls concerning food additives in EU countries. Three further detailed directives covering authorisation and conditions of use for sweeteners, colours and food additives other than colours and sweeteners (commonly called miscellaneous additives), form a 'comprehensive' directive that controls the specific use in food of all direct food additives in the EU (except flavourings). These directives result from years of discussions to bring about agreement between the EU member states, whose laws on food additives were previously disparate. Both the additives permitted and the degree of control on their use varied greatly throughout the EU countries, from no positive lists for certain additives, through precise lists with conditions of use to authorisation only with specific permission from the authorities. As expected, countries with strict controls have been unwilling to relax these, so that the resultant directives have veered towards more stringent control of the use of additives in food products. All listed food additives have now been evaluated for their safety and the recommendations of the Scientific Committee for Food are taken into account during considerations for the inclusion and conditions of use in food of every food additive.

This chapter also summarises EU laws applicable to flavourings and extraction solvents (which are not within the scope of Directive 89/107/EEC), and labelling requirements for additives used in food products, including specific requirements relating to the use of packaging gases, sweeteners and genetically modified (GM) additives and flavourings, and for food additives sold as such, both in trade and direct to the consumer. Some practical problems of implementation of the EU laws are addressed, and future developments in the field of additives legislation are outlined.

4.2 The key directives

4.2.1 Directive 89/107/EEC

Directive 89/107/EEC,[1] sometimes known as a 'framework' directive, contains a number of general safety and authorising measures concerning food additives. There is a definition of 'food additive' as follows:

> any substance not normally consumed as a food in itself and not normally used as a characteristic ingredient of food whether or not it has nutritive value, the intentional addition of which to food for a technological purpose in the manufacture, processing, preparation, treatment, packaging, transport or storage of such food results, or may be reasonably expected to result, in it or its by-products becoming directly or indirectly a component of such foods.

The definition does not include processing aids, and these substances therefore are not regarded as food additives. A processing aid is a substance (not consumed as a food ingredient by itself) that is intentionally used for the processing of raw materials, foods or ingredients to fulfil a certain technological purpose during treatment or processing. It may result in unintentional but technically unavoidable residues of the substance or its derivatives in the final product, but these residues must not have any technological effect on the final product, and they must not present any health risk. There is no EU positive list of processing aids. Plant health products, flavourings covered by Directive 88/388/EEC and substances used as nutrients (e.g. vitamins, trace elements, minerals) in foodstuffs are not regarded as additives.

Annex I contains a list of additive categories covered by the additives directives, as shown in Table 4.1. Annex II sets out general criteria concerning technological need for, safety and use of food additives. Food additives may only be approved if a technological need is demonstrated and the purpose cannot be achieved by other technological means, they do not present a hazard to health at the proposed levels of use, and they do not mislead the consumer. Additives may only be used for certain purposes including improvement of keeping quality and stability of food and to aid in the manufacture, processing and treatment of food, when they must not mask the use of faulty raw materials or undesirable practices. They must undergo appropriate toxicological testing and must be re-evaluated when necessary. The approval for food additives must specify the foodstuffs to which additives may be added and the conditions of their use, be limited to the lowest level of use needed to achieve the intended effect and take account of any acceptable daily intake, or equivalent assessment, established for the food additive and the likely intake of it from all sources. Intake of additives by special groups of consumers must also be considered.

Table 4.1 Categories of additives listed in annex I of Directive 89/107/EEC

Colour	Modified starch
Preservative	Sweetener
Antioxidant	Raising agent
Emulsifier	Anti-foaming agent
Emulsifying salt	Glazing agent
Thickener	Flour treatment agent
Gelling agent	Firming agent
Stabiliser	Humectant
Flavour enhancer	Sequestrant
Acid	Enzyme
Acidity regulator	Bulking agent
Anti-caking agent	Propellant gas and packaging gas

The directive makes provision for drawing up rules concerning detailed lists of authorised food additives, conditions and limitations of their use, use of carrier solvents, purity criteria for food additives and carrier solvents, and sampling and analysis of additives in and on foods. These lists must be contained in a 'comprehensive' directive, and this may be drawn up in stages. Provisions for temporary, national marketing of a non-listed additive, and for member states to suspend or restrict use of an authorised additive in their country on grounds of suspected danger to health, are made. There is a specific requirement that provisions affecting public health are referred to the Scientific Committee for Food. Labelling rules for additives sold as such, both in trade and to the consumer, are prescribed. These details are stated in section 4.2.9.

Traditional foods

Directive 94/34/EC,[1] an amendment to Directive 89/107/EEC, allows member states to prohibit in their country certain classes of additives in the production and sale of foodstuffs that they consider as traditional. They must, however, allow the production and sale of all foodstuffs that are considered as non-traditional. The list of 'traditional' foods was published in Commission Decision No. 292/97;[2] it includes, for instance, prohibitions on all additives in Greek 'feta' cheese, all additives except propellant gases in traditional German beer, and all additives except preservatives and antioxidants in Spanish traditional 'Lomo embuchado'.

4.2.2 The 'comprehensive' directive

The three directives forming the 'comprehensive' directive concern sweeteners,[3] colours[4] and all other classes of additives[5] (or miscellaneous additives) permitted in food and drink products. They are usually referred to by their specific names or as the 'specific' directives. These technical directives are of key interest to the food manufacturer since they list the authorised additives, the foods in which these additives may be used and, where necessary, maximum use levels. Some terms are common to all three directives, and these are discussed below.

Expressions of quantity
The annexes to the directives are very specific, listing the permitted additives and in many instances the levels permitted in specified foods. Generally these maximum levels are expressed in mg/kg or mg/l as on the product ready to eat when prepared according to the manufacturers' instructions for use. For colours the quantity refers to the colouring principle contained in any colouring preparation. In the case of the directive concerning all other classes of additives, however, the expressions may refer to the foods as sold. Therefore it is necessary to consider the stage of the process when calculating levels of additives.

The term quantum satis
This term appears frequently in the directives and is used when no maximum level of use is prescribed. Prescription of a maximum quantity is not considered necessary for many additives which are considered safe under envisaged conditions of use. *Quantum satis* means that the additive may be used in food in accordance with good manufacturing practice, at a level not higher than that needed to achieve the intended technological purpose, and provided that the consumer is not misled.

Carry-over and reverse carry-over
The specific directives contain provisions that allow prepared products to contain additives that may not be listed but are present because they were used lawfully and in correct proportions in an ingredient used to prepare the product. This provision is usually known as 'carry-over'. They also make provisions for an intermediate ingredient to contain an additive that would not normally be permitted, if that ingredient is to be used only to prepare a product in which the additive is permitted. This situation is sometimes called 'reverse carry-over'. The miscellaneous additives directive does not allow carry-over provisions to apply to infant formulae, follow-on formulae or weaning foods covered by Directive 89/398/EC, except where this is specially provided for.

4.2.3 Sweeteners

Directive 94/35/EC[3] on sweeteners for use in foodstuffs applies to food additives that are used to impart a sweet taste to foodstuffs or used as table-top sweeteners. Most authorisations are for foods with no added sugar or foods that are energy-reduced. The term 'with no added sugar' means without any added mono- or disaccharides or any other foodstuff used for its sweetening properties, and the term 'energy-reduced' means with an energy value reduced by at least 30% compared with the original foodstuff or a similar product. Foods that have sweetening properties (such as honey) are not regarded as sweeteners. Sweeteners are not permitted in food including dietary food (unless otherwise specified) for infants and young children covered by Directive 89/398/EEC. Table 4.2 shows the sweeteners permitted for use in foodstuffs or for sale to the consumer.

Table 4.2 List of permitted sweeteners

E 420	Sorbitol: (i) sorbitol, (ii) sorbitol syrup
E 421	Mannitol
E 953	Isomalt
E 965	Maltitol: (i) maltitol, (ii) maltitol syrup
E 966	Lactitol
E 967	Xylitol
E 950	Acesulfame K
E 951	Aspartame
E 952	Cyclamic acid, sodium and calcium cyclamates (use levels expressed as free acid)
E 954	Saccharin, sodium, potassium and calcium saccharin (use levels expressed as free imide)
E 957	Thaumatin
E 959	Neohesperidine DC

E 420, E 421, E 953, E 965, E 966 and E 967 are polyhydric alcohols and they are allowed to *quantum satis* in: certain foods that are energy-reduced or with no added sugar, including certain dessert products, breakfast cereals, edible ices, jams and marmalades, certain confectionery products, fine bakery wares and sandwich spreads; in chewing gum with no added sugar; and in sauces, mustard, products for particular nutritional uses and solid supplements/dietary integrators.

E 951, E 952, E 954, E 957 and E 959 are intensely sweet substances and conditions of use are imposed, with maximum limits prescribed in certain foods for each sweetener. A few examples of authorised uses of these sweeteners are stated in Table 4.3.

4.2.4 Colours

The directive on colours for use in foodstuffs, 94/36/EC,[4] replaced a previous positive list of colours, but it goes much further to harmonise completely the laws of the EU member countries in respect of food colours, since, for many colours, the foods in which certain colours are permitted and maximum levels of use are prescribed. Colours are defined as: 'substances which add or restore colour in a food, and include natural constituents of foodstuffs and natural sources which are normally not consumed as foodstuffs as such and not normally used as characteristic ingredients of food'.

The definition includes substances obtained from foods and other natural sources by physical and/or chemical extraction that results in selective extraction of the pigments, but it does not include foodstuffs, flavourings that have a secondary colouring effect, such as paprika, turmeric and saffron, or colours used on external inedible parts of foods, such as cheese coatings and sausage casings. The colours that may be used for health marking and other required marking on meat products and for decorative colouring of eggshells are prescribed. The main technical details are contained in five annexes to the directive.

Table 4.3 Examples of uses of authorised sweeteners

Food product	Maximum limit of use					
	E 950	E 951	E 952	E 954	E 957	E 959
Water-based flavoured drinks, energy reduced or with no added sugar	350 mg/l	600 mg/l	400 mg/l	80 mg/l	–	30 mg/l
Confectionery with no added sugar	500 mg/kg	1,000 mg/kg	500 mg/kg	500 mg/kg	50 mg/kg	100 mg/kg
Energy-reduced beer	25 mg/l	25 mg/l	–	–	–	10 mg/kg
Edible ices, energy reduced or with no added sugar	800 mg/kg	800 mg/kg	250 mg/kg	100 mg/kg	50 mg/kg	50 mg/kg
Food supplements/dietary integrators based on vitamins and/or mineral elements, syrup-type or chewable	2,000 mg/kg	5,500 mg/kg	1,250 mg/kg	1,200 mg/kg	400 mg/kg	–
Breakfast cereals with a fibre content more than 15%, containing at least 20% bran, energy reduced or with no added sugar	1,200 mg/kg	1,000 mg/kg	–	100 mg/kg	–	50 mg/kg
Energy-reduced soups	110 mg/l	110 mg/l	–	110 mg/l	–	50 mg/l
Breath-freshening micro-sweets, with no added sugar	2,500 mg/kg	6,000 mg/kg	2,500 mg/kg	3,000 mg/kg	–	400 mg/kg

Annex I contains the list of permitted colours for food. Only colours included on this list may be sold directly to consumers, except that E 123, E 127, E 128, E 154, E 160b, E 161g, E 173 and E 180 may not be sold directly to consumers. The complete list of authorised colours is stated in Table 4.4.

Annex II lists foodstuffs that may not contain added colours unless these are expressly permitted by other annexes or they are present because of legitimate carry-over in an ingredient. The list includes unprocessed foods and processed foods that would not be expected to contain colours, also some processed foods listed in subsequent annexes which may contain only a few colours. The list includes bottled waters, milk, cream, oils and fats, eggs and egg products, flour, bread, pasta, sugar, processed fruit and vegetables, extra jam, coffee and tea and preparations of these, salt, honey, certain spirits, and wine covered by Regulation (EEC) No. 822/87.

Annex III lists foods in which only certain colours are permitted, detailing the specific permitted colours with maximum levels for each food. Examples include margarine to which E 160a and E 100 may be added *quantum satis*, and E 160b max. 10 mg/kg; vinegar and liqueur wines, which may only contain E 150a–d *quantum satis*; breakfast sausages with a minimum cereal content of 6% and burger meat with a minimum vegetable and/or cereal content of 4%, which may contain E 129 max. 25 mg/kg, E 120 max. 100 mg/kg and E 150a–d *quantum satis*; dried potato granules which may only contain E 100 *quantum satis*.

Annex IV lists colours with very restricted food use: E 123, E 127, E 128, E 154, E 161g, E 173, E 174, E 175, E 180 and E 160b. Examples of permitted uses include E 123 which may be added to aperitif wines, spirit drinks including products with less than 15% alcohol by volume, max. 30 mg/l, and fish roe, max. 30 mg/kg; E 154 which may be added to kippers, max. 20 mg/kg; E 174 and E 175 which may be used for external coating of confectionery, decoration of chocolates and in liqueurs, *quantum satis*; and E 180 which may be used for edible cheese rind, *quantum satis*.

Table 4.4 List of authorised colours (annex 1 to Directive 94/36/EC)[1]

E number	Colour	Colour Index number
E 100	Curcumin	75300
E 101	(i) Riboflavin (ii) Riboflavin-5'phosphate	
E 102	Tartrazine	19140
E 104	Quinoline Yellow	47005
E 110	Sunset Yellow FCF Orange Yellow S	15985
E 120	Cochineal, Carminic acid, Carmines	75470
E 122	Azorubine, Carmoisine	14720
E 123	Amaranth	16185
E 124	Ponceau 4R, Cochineal Red A	16255

E number	Colour	Colour Index number
E 127	Erythrosine	45430
E 128	Red 2G	18050
E 129	Allura Red AC	16035
E 131	Patent Blue V	42051
E 132	Indigotine, Indigo carmine	73015
E 133	Brilliant Blue FCF	42090
E 140	Chlorophylls and Chlorophyllins: (i) Chlorophylls (ii) Chlorophyllins	75810 75815
E 141	Copper complexes of chlorophylls and chlorophyllins (i) Copper complexes of chlorophylls (ii) Copper complexes of chlorophyllins	75815
E 142	Green S	44090
E 150a	Plain caramel[2]	
E 150b	Caustic sulphite caramel	
E 150c	Ammonia caramel	
E 150d	Sulphite ammonia caramel	
E 151	Brilliant Black BN, Black PN	28440
E 153	Vegetable carbon	
E 154	Brown FK	
E 155	Brown HT	20285
E 160a	Carotenes: (i) Mixed carotenes (ii) Beta-carotene	 75130 40800
E 160b	Annatto, bixin, norbixin	75120
E 160c	Paprika extract, capsanthin, capsorubin	
E 160d	Lycopene	
E 160e	Beta-apo-8'-carotenal (C30)	40820
E 160f	Ethyl ester of beta-apo-8'-carotenic acid (C30)	40825
E 161b	Lutein	
E 161g	Canthaxanthin	
E 162	Beetroot Red, betanin	
E 163	Anthocyanins	Prepared by physical means from fruits and vegetables
E 170	Calcium carbonate	77220
E 171	Titanium dioxide	77891
E 172	Iron oxides and hydroxides	77491, 77492 77499
E 173	Aluminium	
E 174	Silver	
E 175	Gold	
E 180	Litholrubine BK	

Notes: [1]Aluminium lakes of listed colours are also permitted.
[2]Caramel means the brown products intended for colouring, not the sugary product obtained by heating sugars which is used for flavouring food such as confectionery, pastry and alcoholic drinks.

Annex V is the most complicated annex and lists the most common uses of colours in foodstuffs. Part 1 contains a list of colours, mostly of natural derivation, that may be added generally to foods at *quantum satis*, unless this is prohibited by earlier annexes. These are: E 101, E 140, E 141, E 150a–d, E 153, E 160a, E 160c, E 162, E 163, E 170, E 171 and E 172. Part 2 contains a list of colours that may be used, singly or in combination, to the maximum level in foods specified in an accompanying table. However, amounts of each of the colours E 110, E 122, E 124, and E 155 may not exceed 50 mg/kg or mg/l in non-alcoholic flavoured drinks, edible ices, desserts, fine bakery wares and confectionery. The part 2 list of colours is: E 100, E 102, E 104, E 110, E 120, E 122, E 124, E 129, E 131, E 132, E 133, E 142, E 151, E 155, E 160d, E 160e, E 160f and E 161b. The foods listed in the table to this annex include non-alcoholic flavoured drinks max. 100 mg/l; edible ices max. 150 mg/kg; desserts max. 150 mg/kg; smoked fish max. 100 mg/kg; dry, savoury potato, cereal or starch-based snack products (extruded or expanded) max. 200 mg/kg, (others) max. 100 mg/kg; liquid food supplements/dietary integrators max. 100 mg/l; solid food supplements/dietary integrators max. 300 mg/kg; spirituous beverages, aromatised wines and similar products (unless mentioned in annexes II or III), fruit wines, cider, perry max. 200 mg/l.

4.2.5 Food additives other than colours and sweeteners

Directive 95/2/EC, as amended by Directives 96/85/EC, 98/72/EC and 2001/5/EC,[5] contains the rules for use in food of other classes of food additives, except flavourings. This directive is the most complicated of the specific directives, and makes provisions for the use of a large number of food additives, with varying degrees of control. There is a long list of additives that may be used to *quantum satis*, lists of antioxidants and preservatives and several other additives that are very strictly controlled, a list of foods in which only certain additives may be used and lists of foods for infants and young children that may contain certain additives. The provisions of the directive apply generally to corresponding foods intended for particular nutritional uses covered by Directive 89/398/EEC except such foods for infants and young children where specific authorisations are made. The directive does not apply to enzymes, except invertase and lysozyme.

The categories of additives covered by Directive 95/2/EC are: preservatives, antioxidants, carriers including carrier solvents, acids, acidity regulators, anti-caking agents, anti-foaming agents, bulking agents, emulsifiers, emulsifying salts, firming agents, flavour enhancers, foaming agents, gelling agents, glazing agents, humectants, chemically modified starches, packaging gases, propellants, raising agents, sequestrants, stabilisers and thickeners. Each of these categories is defined. Substances that are not regarded as additives for the purposes of this directive are also listed: substances used for treatment of drinking water, liquid pectin, chewing gum bases, dextrins, starches and physically modified starches (however, chemically modified starches are regarded as food additives), ammonium chloride, blood plasma, protein hydrolysates, gelatin, milk protein, gluten, most amino acids that have no additive function, casein and caseinates and inulin. Six annexes are appended to the directive, and these contain the major technical provisions.

Annex I contains a long list of additives permitted generally in food products to *quantum satis*. These additives include acetic acid and its potassium, sodium and calcium salts (E 260, E 261, E 262), ascorbic acid and its sodium and calcium salts (E 300, E 301, E 302), citric acid and its sodium, potassium and calcium salts (E 330, E 331, E 332, E 333), alginic acid and alginates (E 400–E 404), gums, pectins, fatty acids and salts and esters of fatty acids, sodium, potassium, ammonium and magnesium carbonates, glucono-delta-lactone and sodium, potassium and calcium gluconates, the packaging gases argon, helium, nitrogen, nitrous oxide and oxygen, polydextrose and chemically modified starches. These additives may not be used in unprocessed foods, honey, non-emulsified animal or vegetable oils, butter, milk, cream, fermented milk products, natural mineral water, spring water, coffee and coffee extracts, unflavoured leaf tea, sugars, pasta and natural unflavoured buttermilk, or in the foods listed in annex II or in foods for infants and young children, unless specific provisions are made in other annexes for such use. However, carbon dioxide and other packaging gases may be used in these foods.

Annex II lists foodstuffs for which only certain of the annex I additives may be used. Such foodstuffs include cocoa products and chocolate products, fruit juices and nectars, jam, jellies and marmalades and partially dehydrated and dehydrated milk, which are the subjects of EU vertical standards, and a number of other foods including frozen unprocessed fruit and vegetables, quick-cook rice, non-emulsified oils and fats, canned and bottled fruit and vegetables, bread made with basic ingredients only, fresh pasta and beer.

Annex III lays down the conditions of use of permitted preservatives and antioxidants, with lists of foods and maximum levels in each case. Part A lists the sorbates, benzoates and p-hydroxybenzoates, E 200–E 219; part B lists sulphur dioxide and the sulphites, E 220–E 228; part C lists other preservatives with their uses, including nisin, dimethyl dicarbonate and substances allowed for surface treatment of certain fruits, E 249 potassium nitrite, E 250 sodium nitrite, E 251 sodium nitrate and E 252 potassium nitrate, E 280–E 283 propionic acid and the propionates; part D lists the antioxidants E 320 butylated hydroxyanisole (BHA), E 321 butylated hydroxytoluene (BHT), E 310 propyl gallate, E 311 octyl gallate, E 312 dodecyl gallate, E 315 erythorbic acid and E 316 sodium erythorbate.

Annex IV contains a long list of other additives that are permitted with restrictions in named foods. This list includes E 297 fumaric acid, phosphates (E 338, E 339, E 340, E 341, E 343, E 450, E 451 and E 452), E 405 propane-1,2-diol alginate, E 416 karaya gum, sorbitol and other polyols (when not used for sweetening purposes), polysorbates (E 432–E 436), E 473 sucrose esters of fatty acids and E 474 sucroglycerides, E 481 sodium stearoyl-2-lactylate and E 482 calcium stearoyl-2-lactylate, (E 491–E 495) sorbitan esters of fatty acids, E 551–E 559 silicon dioxide and certain silicates, (E 620–E 635) glutamic acid, glutamates, guanylic acid and guanylates, inosinic acid and inosinates, ribonucleotides, and acesulfame-K, aspartame, thaumatin and neohesperidine DC when used as flavour enhancers, E 999 quillaia extract and E 1505 triethyl citrate.

Annex V lists permitted carriers and carrier solvents, some of which may only be used for restricted purposes.

Annex VI lists the additives permitted in infant foods and foods for young children. Part 1 lists the few additives allowed in infant formulae for infants in good health, part 2 those allowed in follow-on formulae for infants in good health, and part 3 the additives permitted in weaning foods for infants and young children in good health. Part 4 applies the lists in parts 1–3 to foods for infants and young children for special medical purposes.

It is to be expected that a directive that covers so many additives would be amended frequently, and the two amendments already published take account of technical developments in the field of food additives since the original Directive 95/2/EC was adopted. Directive 96/85/EC adds E 407a processed eucheuma seaweed to the annex I list of generally permitted additives. Directive 98/72/EC includes E 920 L-cysteine in the annex I list of generally permitted additives, but only for use as a flour treatment agent. E 469 enzymatically hydrolysed carboxy methyl cellulose, E 1103 invertase and E 1451 acetylated oxidised starch are also added to annex I. The further annexes are also amended by the addition of several new additives and extended uses of existing additives, with conditions of use laid down. Further authorisations are listed in annex VI, including a list of additives permitted with restrictions in specific foods for infants and young children for special medical purposes.

Amending Directive 2001/5/EC contains several new authorisations including the addition of E 949 hydrogen to annex I, and E 943a butane, E 943b iso-butane and E 944 propane to annex IV for use as vegetable oil pan sprays (professional use only) and as water-based emulsion sprays, *quantum satis*. E 650 zinc acetate is also added to annex IV for use in chewing gum, max. 1000 mg/kg.

4.2.6 Purity criteria for additives

As required by Directive 89/107/EEC, criteria of purity have been drawn up for most food additives, and those remaining are scheduled for early completion. Purity criteria for all the permitted sweeteners have been prescribed in Directive 95/31/EC, as amended by Directive 98/66/EC,[6] and criteria for all the permitted colours are contained in Directive 95/45/EC, as amended by Directive 1999/75/EC.[7] Directives that prescribe purity criteria for all the additives authorised under Directive 95/2/EC have been drawn up in stages. Directive 96/77/EC containing purity criteria for antioxidants and preservatives is amended by Directives 98/86/EC which lays down purity criteria for emulsifiers, stabilisers and thickeners and 2000/63/EC[8] which contains putty criteria for most additives numbered E 500 and above, and for certain other additives not covered in the earlier directives. Purity criteria for the few remaining permitted miscellaneous additives will be included in a further amendment to directive 96/77/EC which was agreed by the Standing Committee for Foodstuffs on 15 February 2001; however, purity criteria for E 1201 polyvinylpyrrolidone and E 1202 polyvinylpyrrolidone are still being considered by the Scientific Committee on Food. Some methods of analysis for verifying prescribed purity criteria have been developed at EU level; these are contained in Directive 81/712/EEC.[9]

4.2.7 Flavourings

Harmonisation of laws relating to flavourings used in food production was considered before Directive 89/107/EEC[1] and the subsequent specific directives concerning additives were drawn up. Directive 88/388/EEC[10] relates to flavourings and to source materials for their preparation. It is a framework directive which prescribes some general requirements for the food use of flavourings, and requires that further appropriate provisions are drawn up, including further laws regarding various types of flavourings, additives needed for production, use and storage of flavourings and procedures for sampling and analysis of flavourings in or on foods.

The directive does not apply to edible substances to be consumed as such, substances with exclusively a sweet, sour or salt taste or to material of vegetable or animal origin that has flavouring properties but is not used as a flavouring source. It includes definitions of 'flavouring', 'flavouring substance', 'flavouring preparation', 'process flavouring' and 'smoke flavouring'.

A 'flavouring' is defined as 'flavouring substances, flavouring preparations, process flavourings, smoke flavourings or mixtures'. A 'flavouring substance' means a defined chemical substance with flavouring properties, which is obtained: by physical, enzymatic or microbiological processes from appropriate vegetable or animal material that is either raw or processed for human consumption using traditional food preparation processes (including drying, torrefaction and fermentation); or by chemical synthesis or isolated by chemical process and is chemically identical to a substance naturally present in appropriate vegetable or animal material; or by chemical synthesis but is not chemically identical to a substance naturally present in appropriate vegetable or animal material. A 'flavouring preparation' is a product other than a flavouring substance, concentrated or not, with flavouring properties, which is obtained by physical, enzymatic or microbiological processes from appropriate vegetable or animal material, either raw or processed using traditional food preparation processes. A 'process flavouring' is a product obtained by heating to maximum 180°C for fifteen minutes a mixture of ingredients which may or may not have flavouring properties, of which at least one contains amino nitrogen and another is a reducing sugar. 'Smoke flavouring' is a smoke extract used in traditional smoking processes for foods.

Some general purity criteria are listed, and limits for certain substances in foods resulting from the use of flavourings are prescribed. Annex I prescribes a limit for 3,4 benzopyrene of 0.03 μg/kg in foods and beverages, and annex II details limits for other substances, including beta asarone, coumarin, hydrocyanic acid, pulegone and safrole. Currently, the definition of 'process flavouring' and some of the limits for undesirable substances in foods are being reviewed.

Directive 88/388/EEC, completed by Directive 91/71/EEC,[11] also prescribes labelling requirements for flavouring sold as such, both in trade and to the final consumer. Such flavourings must be designated by the word 'flavouring' or a more specific description, the phrase 'for foodstuffs' or a more specific reference to their intended food use, and the name and address of the manufacturer or packer, or seller in the EU, the nominal quantity and a batch reference must be given. In the case of trade sales the categories of flavouring substances and flavouring preparations present and of any other components must be listed together with the quantity of any component that is subject to a quantitative limit in the foodstuff. Some of this information may be placed on trade documents supplied in advance or with the consignment. For sales to consumers the label information must include the names of any other substances present, the date of minimum durability, special conditions for storage and use, if necessary, and instructions for use, if needed. For all sales of flavourings where the type of flavouring is not specified, the word 'natural' may only be used to describe the flavouring if it is derived from natural flavouring substances and/or flavouring preparations only. If the name of the flavouring includes a foodstuff or a flavouring source the word 'natural' may only be used if the flavouring component has been obtained, by means of appropriate physical, enzymatic or microbiological processes or traditional food preparation processes, solely or almost solely from the foodstuff or flavouring component concerned.

Council Decision 88/389/EEC[12] required the Commission to draw up an inventory of source materials and substances used in the preparation of flavourings, and to update this inventory regularly. Regulation 2232/96,[13] the first specific measure drawn up under Directive 88/388/EEC, sets out a staged procedure whereby an exclusive list of chemical flavouring substances for food use is to be drawn up. It applies to the following types of flavouring substances: those obtained by physical, enzymatic or microbiological processes from vegetable or animal raw materials; chemically synthesised or chemically isolated flavouring substances that are chemically identical to flavouring substances naturally present in foods or in herbs and spices normally considered as foods; chemically synthesised or chemically isolated flavouring substances that are chemically identical to flavouring substances naturally present in vegetable or animal raw materials that are not normally considered as foods; other chemically synthesised or chemically isolated flavouring substances. The annex to the regulation prescribes safety and use criteria for these flavouring substances.

The regulation contains details of the procedure for drawing up a positive list of flavouring substances. Member states must first notify the Commission of a list of such flavouring substances that might be used in or on foods marketed nationally; the notification should include relevant information such as the nature of the flavouring substance (chemical formula, CAS number, EINECS number, IUPAC classification, etc.), the foods in or on which the flavouring is used and the level of compliance with the limits prescribed in Directive 88/388/EEC for certain undesirable substances in flavourings. On the basis of this information the

Commission must draw up a register of notified flavouring substances, the legal use of which in one member state must be recognised by the other member states. This part of the procedure has been completed and the register is contained in Commission Decision 1999/217/EC.[14] An evaluation programme must then be adopted and this must define in particular the order of priorities for evaluation of the flavouring substances taking account of their uses, the time limits and the flavouring substances that are to be the subject of scientific cooperation. Several substances have been selected for priority evaluation (including caffeine). The positive list of flavourings authorised to the exclusion of all others should be drawn up within five years of adoption of the evaluation programme.

The concerns of the flavour manufacturing industry with respect to confidentiality of the information divulged to the Commission have been recognised and Regulation 2332/96 requires that the intellectual property rights of the manufacturer are protected. Further rules concerning the handling of confidential information by the Commission and by the authorities in member states and European Economic Area (EEA) states are contained in Commission Communication 98/C131/03[15] and Commission Recommendation 98/282/EC.[16]

4.2.8 Extraction solvents

The extraction solvents that are authorised for use in food production in the EU are not covered by Directive 89/107/EEC,[1] since harmonising legislation was in force before that directive was drawn up. Directive 88/344/EEC, as amended,[17] concerning extraction solvents used in the production of food and food ingredients, was made to harmonise laws in this area with a view to free movement of foodstuffs, taking account of issues concerning public health, economic and technical needs. An extraction solvent means a solvent that is used in an extraction procedure during the processing of raw materials, foodstuffs or of components or ingredients of these, and which is removed but may result in the unintentional but technically unavoidable presence of residues or derivatives in the final food or ingredients. Water, foods with solvent properties and ethanol may be used as extraction solvents generally in the production of foods and food ingredients.

The annex to Directive 84/344/EEC, as amended, which contains the technical details, includes three lists of extraction solvents allowed for food processing: those permitted for all uses according to good manufacturing practice; extraction solvents for which conditions of use are specified; and extraction solvents for the preparation of flavourings, with conditions of use specified. The directive has been amended three times to take account of the recommendations of the Scientific Committee for Food and the need for certain solvents by the food industry. Currently the permitted lists of extraction solvents are as follows:

- Part I: extraction solvents allowed for all uses according to good manufacturing practice: propane, butane, ethyl acetate, ethanol, carbon dioxide, acetone (not for production of olive-pomace oil), nitrous oxide.

- Part II: extraction solvents for which conditions of use/maximum residue limits, are specified: hexane, methanol, propan-2-ol, methyl acetate, ethylmethylketone, dichloromethane. Common uses include fractionation of oils and fats, preparation of defatted products and decaffeination for coffee and tea.
- Part III: extraction solvents that may be used for the preparation of flavourings from natural flavouring materials: diethyl ether, hexane, methyl acetate, butan-1-ol, butan-2-ol, ethylmethyl ketone, dichloromethane, propan-1-ol, cyclohexane, 1,1,1,2-tetrafluoroethane. Maximum residue limits in the final food are stated for each solvent. In parts II and III the combined use of hexane and ethylmethyl ketone is not allowed.

General criteria of purity are prescribed for the listed extraction solvents. Extraction solvents need not be listed in the ingredients list of food products. When sold as such for business purposes the packaging or containers must be labelled to show the correct name of the extraction solvent, the name and address of the responsible person, an indication of the suitability of the solvent for food use, the batch or lot reference, the net volume and any special conditions for storage and use. Some of this information may be given on trade documents supplied with or prior to delivery of the extraction solvent.

4.2.9 Labelling requirements for additives

General requirements for labelling of additives used as food ingredients

Generally food additives that are used in the manufacture of food products must be identified as prescribed in the Labelling Directive 2000/13/EC,[18] (formerly 79/112/EEC, as amended). The correct category name, representing the function of the additive in the food, must be stated, followed by the specific name or E number of the additive or additives present. If an additive performs more than one function, the category that represents its principal function in that food must be named. The list of category names is provided in Table 4.5.

The E number need not be stated in the case of modified starches; however, if a modified starch could contain gluten, the category name must be accompanied by an indication of the specific vegetable origin of the starch. The category name 'emulsifying salt' may only be used for processed cheese and processed cheese products. An example of a correctly identified additive could be 'preservative – sodium benzoate' or 'preservative E 211'. The indications of additives must be placed in the list of ingredients of the product in the correct position in order by weight (greatest first).

There are certain exemptions from these rules. Some food ingredients need only be identified by a generic term, and additives used in such ingredients need not be named. Additives contained in ingredients of a food need not be listed, provided that they do not perform a technological function in the final foodstuff, and additives used as processing aids need not be listed. The constituents of compound ingredients that constitute less than 25% of the product need not be stated. Additives present in a compound ingredient must usually be stated as

Table 4.5 Permitted category names for food additives in food labelling

Acid	Flour treatment agent
Acidity regulator	Gelling agent
Anti-caking agent	Glazing agent
Anti-foaming agent	Humectant
Antioxidant	Modified starch
Bulking agent	Preservative
Colour	Propellant gas
Emulsifier	Raising agent
Emulsifying salts	Stabiliser
Firming agent	Sweetener
Flavour enhancer	Thickener

described above, but if such an additive does not perform a significant technological function in the final product, it need not be identified in the ingredients list. However, where a flour treatment agent has been used for bread manufacture, it must always be identified, in the form described above.

Flavourings are not included in the categories listed above. When used in food products they must be designated by the term 'flavourings', or by a more specific name or description of the flavouring. Flavourings may be described as 'natural' only if the flavouring component consists solely of flavouring preparations and/or flavouring substances as defined in Directive 88/388/EEC.[9] If the name of the flavouring refers to the vegetable or animal nature or origin of the substances contained in it, the word 'natural' or a similar term may only be used if the flavouring component has been isolated by appropriate physical, enzymatic or microbiological processes or traditional food preparation processes, solely or almost solely from that vegetable or animal source.

Specific requirements for identification of certain additives in foods

Packaging gases

Considerations of the need to provide consumers with adequate information about the food products they buy have played a large part in the developments of increasingly detailed labelling rules. Most additives must be identified as described above, but this requirement does not apply to packaging gases. Nevertheless, consumers should be aware when these substances are used to prolong the shelf-life of the food. Directive 94/54/EC[19] states that where the durability of a foodstuff has been extended by means of packaging gases (as permitted), the label should bear the statement 'packaged in a protective atmosphere'.

Sweeteners/sugars, aspartame, polyols

Table-top sweeteners must be named '. . . -based table-top sweetener', including the name(s) of the sweetener(s) used.

Directive 94/54/EC has been amended by Directive 96/21/EC[19] which includes further specific requirements relating to the presence of sweeteners and warnings in respect of certain sweeteners. The statements, which must be made in addition to the declarations required in the list of ingredients, are detailed as follows.

Where a foodstuff contains a sweetener or sweeteners, as allowed, the statement 'with sweetener/s' must accompany the name of the product. Where a foodstuff contains both added sugar and sweeteners, as allowed, the statement 'with sugar/s and sweetener/s' must accompany the name of the product. Foodstuffs that contain aspartame must bear the statement 'contains a source of phenylalanine'. Foodstuffs that contain more than 10% added polyols must bear the statement 'excessive consumption may produce laxative effects'.

Labelling requirements for additives sold as such

Requirements for the labelling of additives for business and consumer sales are contained in articles 7 and 8 of Directive 89/107/EEC.[1] For business sales, packaging or containers must bear the following information: the name and E number required by law or, if none, a precise description of the additive/s, in descending order of proportion by weight; if other substances are present to facilitate storage sale, standardisation, dilution or dissolution of the additive, an indication of each component in descending order of proportion by weight in the product; a statement 'for use in food' or 'restricted use in food' or a specific reference to its food use; any special conditions of storage and use; directions for use, if necessary; a batch or lot mark; the name or business name and address of the responsible person in the EU; where a component or group of components has a quantitative limit in foods the percentage present or other information that enables the purchaser to comply with the relevant law; the net quantity. The information must be conspicuous, clearly legible and indelible. Some of it may appear on relevant documents supplied with or prior to the delivery, but the name of the additive and the statements relating to food use must be placed on the labelling of the packaging or container.

In the case of sales to the consumer the packaging must bear the name and E number or, if none, a precise description of the product, and the information stated for business sales above (except that relating to the percentage of components present). The minimum durability of the product must be stated, according to the requirements detailed in Directive 2000/13/EC.[18]

4.2.10 Practical implementation of the EU additives directives

Directives must be implemented nationally before they have legal force; the directives concerning food additives have been implemented into the food law of all the existing EU member states. In general this means that trade problems in this area should be eliminated, but the manufacturer needs to be aware of difficulties that may still remain. Some of these are summarised below.

1. Countries must implement the EU directives, but there is flexibility as to how this is accomplished, provided that the technical requirements are met. In the UK regulations have been made on sweeteners,[20] colours,[21] miscellaneous additives,[22] flavourings,[23] extraction solvents[24] and labelling of food

additives when sold as such[25] and when used in foodstuffs,[26, 27] and these adopt the EU laws, including the relevant technical requirements of Directive 89/107/EEC and references to criteria of purity for the relevant additives. The annexes to the directives have been transmitted to the regulations in much the same way as they are printed in the EU directives. The UK government has also produced non-statutory guidance notes on the additives laws generally and on the rules concerning sweeteners, colours and miscellaneous additives used in foods.[28] However, in other member states the implementation takes different forms, with many countries listing food groups and the additives that are permitted in the various foods. In practice the manufacturer should always check the national laws as well as the EU laws, and should be prepared for varying formats of national regulations, and possibly different interpretations of the EU requirements. Enforcement of regulations is also national, in general, and the manufacturer or importer/exporter should become familiar with enforcement practices in the various countries. In fact, it is recommended that those responsible should have some knowledge of the whole food law area, its operation and enforcement in the country in which they are interested in marketing food products.

2. Amendments to directives usually allow a transitional time before the new or amended requirements must be implemented, and this may vary between the member states. Generally, reference can be made to the original directive, but some countries can enforce their existing requirements quite strictly.
3. When trade with Central and Eastern European countries that are looking to accede to the EU is considered, their laws must be considered very carefully, since they are rapidly harmonising them with the EU requirements. It would be necessary to ascertain exactly which laws are in force, and which will remain in force in the future. Similar considerations apply to other European countries, and countries nearby who will be considering changes to their legislation with a view to future accession to the EU and/or elimination of technical trade barriers with EU countries.
4. Manufacturers should seek to keep up to date with food law developments on a global basis, including the changing views and developments in the EU, the USA and Codex Alimentarius. As an example, safety considerations could result in alterations to permitted levels of an additive or even its removal from the authorised lists when producers would need to remove such an additive from products at short notice.

4.3 Requirements contained in vertical food directives

Several directives contain harmonised compositional requirements for specific foods. Originally these included authorisation of permitted additives in such foods, but the specific directives concerning sweeteners, colours and other classes of food additives have now effectively transferred these provisions to the additives framework law. Generally, therefore, the EU controls on food additives are contained in the horizontal legislation described above. However, a few requirements may remain in the 'vertical' directives, and manufacturers should always check the 'vertical' directives that relate to their products. As an example, where jams and similar products contain a residual sulphur dioxide content of more than 30 mg/kg this must be stated in the list of ingredients as percentage by weight or residue in the product as 'sulphur dioxide', in addition to the declaration of the additive in the list of ingredients.

4.3.1 Foods and ingredients that contain genetically modified additives and flavourings

EU Regulation No. 50/2000[29] prescribes the conditions under which foods and food ingredients (for consumer sale or sale to mass caterers) that contain GM additives and flavourings must be identified. The required terms are 'produced from genetically modified ...' in parentheses immediately after the indication in the ingredients list of the additive or flavouring in question or, if appropriate, 'genetically modified ...' immediately after the indication in the ingredients list of the relevant additive or flavouring. Alternatively, the phrases may appear in a prominent asterisked footnote to the ingredients list. The fount of the footnote must be at least the same size as that used for the list of ingredients. For such foods where no list of ingredients is required, the wording quoted above must appear prominently on the label.

4.4 Future trends

As indicated above, proposed and ongoing legislation should be completed in the short term. Amendments to the Directives that authorise the use of food additives, in particular Directive 95/2/EC, will need to be made as fresh requests and concerns are addressed. The few purity criteria for additives other than sweeteners and colours will be completed shortly. The evaluation procedure for flavours listed in the register of flavouring substances will be completed within a few years, and the exclusive positive list of flavourings for food use will be drawn up.

It is likely that separate rules will be drawn up to cover GM additives and flavourings that are sold as such. The Commission is also examining the adventitious contamination of additives and flavourings with DNA or protein resulting from genetic modification, with a view to the possibility of setting a threshold that will avoid labelling as a result of such contamination.

In the longer term there is no doubt that additive laws will continue to be revised and change. Such changes result in large part from increased concerns about food safety and demands and desires for more transparent food laws that provide the consumer with better information about food safety and risks. A proposed Council Regulation[30] contains a definition of food which will include 'any substance intentionally incorporated into the food during its manufacture, preparation or treatment', thus including food additives as food, processed or unprocessed. The Regulation will set out numerous principles of food law and food safety, many of which are not dissimilar to existing food laws in the Member States. A European Food Authority will be set up that will contribute to a high level of protection of human life and health and facilitate the functioning of the internal market, setting up a system of scientific and technical support for the Community's legislation. There will be a Scientific Panel on food additives, flavourings, processing aids and materials in contact with food which will provide opinions that fall within its sphere of competence. The proposed regulation was preceded by an EU White Paper on Food Safety[31] that set out the radical new approach to guarantee a high level of food safety. The White Paper also addressed a number of detailed issues that need to be considered for completion and evaluation and included the following recommendations in respect of food additives:

- Directive 89/107/EEC should be amended to confer implementing powers on the Commission to maintain lists of authorised additives and lay down specifications in respect of enzymes
- the EU lists of sweeteners, colours and other food additives need to be updated
- purity criteria for sweeteners, colours and other food additives need to be updated and completed, and for novel additives appropriate purity criteria and a requirement for a new safety evaluation need to be laid down
- further work is needed to reflect innovation in the field of flavourings and investigate toxicological effects of natural substances in flavourings
- the register of flavourings will be updated, in the programme for evaluation of flavourings priorities and time limits will be laid down, a list of additives authorised for use in flavourings will be drawn up and conditions for the production of smoke flavourings prescribed
- the detailed methods of analysis laid down in Directive 81/712/EEC will be replaced with a set of general principles
- in the context of provision for more informative labelling, the labelling of carry-over additives needs to be considered, and also the indication of the presence of food ingredients that are known allergens
- provisions relating to additives produced by genetic engineering will be clarified, and the labelling of food ingredients produced without genetic modification ('GM-free') will be considered.

4.5 Sources of further information and advice

Food Standards Agency
Food Additives Unit
Aviation House
125 Kingsway
London WC2B 6NH

Leatherhead Food Research Association
Randalls Road
Leatherhead
Surrey KT22 7RY

Food Additives and Ingredients Association
Executive Secretary
10 Whitchurch Close
Maidstone
Kent ME16 8UR

4.6 References

1. Council Directive 89/107/EEC of 21 December 1988 on the approximation of the laws of the member states concerning food additives authorised for use in foodstuffs intended for human consumption (*Official Journal of the European Communities* (L40) of 11 February 1989, pp. 27–33), as amended by Directive 94/34/EC of 30 June 1994 (*OJ* (L237) of 10 September 1994, pp. 1–2).
2. Decision No. 292/97/EC of the European Parliament and of the Council of 19 December 1996 on the maintenance of national laws prohibiting the use of certain additives in the production of certain specific foodstuffs (*Official Journal of the European Communities* (L48) of 19 February 1997, pp. 13–15).
3. European Parliament and Council Directive 94/35/EC of 30 June 1994 on sweeteners for use in foodstuffs (*Official Journal of the European Communities* (L237) of 10 September 1994, pp. 3–12), as amended by Directive 96/83/EC of 19 December 1996 (*OJ* (L48) of 19 February 1997, pp. 16–19).
4. European Parliament and Council Directive 94/36/EC of 30 June 1994 on colours for use in foodstuffs (*Official Journal of the European Communities* (L237) of 10 September 1994, pp. 13–29).
5. European Parliament and Council Directive 95/2/EC of 20 February 1995 on food additives other than colours and sweeteners (*Official Journal of the European Communities* (L61) of 18 March 1995, pp. 1–40, as amended by Directives 96/85/EC of 19 December 1996 (*OJ* (L86) of 28 March 1997, p. 4), 98/72/EC of 15 October 1998 (*OJ* (L295) of 4 November 1998, pp. 18–30) and 2001/5/EC of 12 February 2001 (*OJ* (L55) of 24 February 2001, pp. 59–61).
6. Commission Directive 95/31/EC of 5 July 1995 laying down specific criteria of purity concerning sweeteners for use in foodstuffs (*Official Journal of the European Communities* (L178) of 28 July 1995, pp. 1–19), as amended by Directive 98/66/EC of 4 September 1998 (*OJ* (L257) of 19 September 1998, pp. 35–6).

7. Commission Directive 95/45/EC of 26 July 1995 laying down specific purity criteria concerning colours for use in foodstuffs (*Official Journal of the European Communities* (L226) of 22 September 1995, pp. 1–45), as amended by Directive 1999/75/EC of 22 July 1999 (*OJ* (L206) of 5 August 1999, pp. 19–21).
8. Commission Directive 96/77/EC of 2 December 1996 laying down specific purity criteria on food additives other than colours and sweeteners (*Official Journal of the European Communities* (L339) of 30 December 1996, pp. 1–69), as amended by Directives 98/86/EC of 11 November 1998 (*OJ* (L334) of 9 December 1998, pp. 1–63) and 2000/63/EC of 5 October 2000 (*OJ* (L277) of 30 October 2000, pp. 1–61).
9. First Commission Directive 81/712/EEC of 18 July 1981 laying down Community methods of analysis for verifying that certain additives used in foodstuffs satisfy criteria of purity (*Official Journal of the European Communities* 1981 (L257), pp. 1–27).
10. Council Directive 88/388/EEC of 22 June 1988 on the approximation of the laws of the member states relating to flavourings for use in foodstuffs and to source materials for their production (*Official Journal of the European Communities* (L184) of 15 July 1988, pp. 61–6).
11. Commission Directive 91/71/EEC of 16 January 1991 completing Council Directive 88/388/EEC on the approximation of the laws of the member states relating to flavourings for use in foodstuffs and to source materials for their production (*Official Journal of the European Communities* (L42) of 15 February 1991, pp. 25–6).
12. Council Decision 88/389/EEC of 22 June 1988 on the establishment, by the Commission, of an inventory of the source materials and substances used in the preparation of flavourings (*Official Journal of the European Communities* (L184) of 15 July 1988, p. 67).
13. Regulation (EC) No. 2232/96 of the European Parliament and of the Council of 28 October 1996 laying down a Community procedure for flavouring substances used or intended for use in or on foodstuffs (*Official Journal of the European Communities* (L299) of 23 November 1996, pp. 1–4).
14. Commission Decision 1999/217/EC of 23 February 1999 adopting a register of flavouring substances used in or on foodstuffs drawn up in application of Regulation (EC) No. 2232/96 of the European Parliament and of the Council of 28 October 1996 (*Official Journal of the European Communities* (L84) of 27 March 1999, pp. 1–137).
15. Commission Communication 98/C131/03 on the ways in which the Commission is to protect the intellectual property in connection with the development and manufacture of flavouring substances covered by Regulation (EC) No. 2232/96 of the European Parliament and of the Council (*Official Journal of the European Communities* (C131) of 29 April 1998, pp. 3–4).
16. Commission Recommendation 98/282/EC of 21 April 1988 on the ways in which the member states and the signatory states to the Agreement on the

European Economic Area should protect intellectual property in connection with the development and manufacture of flavouring substances referred to in Regulation (EC) No. 2232/96 of the European Parliament and of the Council (*Official Journal of the European Communities* (L127) of 29 April 1998, pp. 32–3).

17. Council Directive 88/344/EEC of 13 June 1988 on the approximation of the laws of the member states on extraction solvents used in the production of foodstuffs and food ingredients (*Official Journal of the European Communities* (L157) of 24 June 1988, pp. 28–33), as amended by Directives 92/115/EEC of 17 December 1992 (*OJ* (L409) of 31 December 1992, pp. 31–2), 94/52/EC of 7 December 1994 (*OJ* (L331) of 21 December 1994, p. 10) and 97/60/EC of 27 October 1997 (*OJ* (L331) of 3 December 1997, pp. 7–9).
18. Directive 2000/13/EC of 20 March 2000 of the European Parliament and of the Council of 20 March 2000 on the approximation of the laws of the member states relating to the labelling, presentation and advertising of foodstuffs (*Official Journal of the European Communities* (L109) of 6 May 2000, pp. 29–42); Corrigendum to Directive 2000/13/EC (*OJ* (L124) of 25 May 2000, p. 66).
19. Commission Directive 94/54/EC of 18 November 1994 concerning the compulsory indication on the labelling of certain foodstuffs of particulars other than those provided for in Council Directive 79/112/EEC (*Official Journal of the European Communities* (L300) of 23 November 1994, pp. 14–15), as amended by Directive 96/21/EC of 29 March 1996 (*OJ* (L88) of 5 April 1996, pp. 5–6).
20. The Sweeteners in Food Regulations 1995 (SI 1995 No. 3123), as amended by SI 1996 No. 1477, SI 1997 No. 814 and SI 1999 No. 982.
21. The Colours in Food Regulations 1995 (SI 1995 No. 3124), as amended by SI 2000 No. 481.
22. The Miscellaneous Food Additives Regulations 1995 (SI 1995 No. 3187), as amended by SI 1997 No. 1413, SI 1999 No. 1136, and SI 2001 No. 60.
23. The Flavourings in Food Regulations 1992 (SI 1992 No. 1971, as amended by SI 1994 No. 1486).
24. The Extraction Solvents in Food Regulations 1993 (SI 1993 No. 1658), as amended by SI 1995 No. 1440 and SI 1998 No. 2257.
25. The Food Additives Labelling Regulations 1992 (SI 1992 No. 1978).
26. The Food Labelling Regulations 1996 (SI 1996, No. 1499).
27. The Genetically Modified and Novel Foods (Labelling) (England) Regulations 2000 (SI 2000 No. 768).
28. Guidance Notes entitled: *New Food Additives Legislation and You*, reference number PB 2522; *Sweeteners in Food Regulations 1995 and You* (revised 1997), reference number PB 2519; *Colours in Food Regulations 1995 and You*, reference number PB 2521; *Miscellaneous Food Additives Regulations 1995 and You*, reference number PB 2520.
29. Commission Regulation (EC) No. 50/2000 of 10 January 2000 on the

labelling of foodstuffs and food ingredients containing additives and flavourings that have been genetically modified or have been produced from genetically modified organisms (*Official Journal of the European Communities* (L6) of 11 January 2000, pp. 15–17).
30. Proposal for a regulation of the European Parliament and of the Council laying down the general principles and requirements of food law, establishing the European Food Authority, and laying down procedures in matters of food, COM (2000) 716 final – 2000/0286(COD) (*Official Journal of the European Communities* (C96 E) of 27 March 2001, pp. 247–68).
31. Commission of the European Communities White Paper on Food Safety, COM (1999) 719 final, Brussels, 12 January 2000.

Appendix: list of E numbers of permitted additives

E 100	Curcumin
E 101	(i) Riboflavin
	(ii) Riboflavin-5'-phosphate
E 102	Tartrazine
E 104	Quinoline Yellow
E 110	Sunset Yellow FCF, Orange Yellow S
E 120	Cochineal, carminic acid, carmines
E 122	Azorubine, carmoisine
E 123	Amaranth
E 124	Ponceau 4R, Cochineal Red A
E 127	Erythrosine
E 128	Red 2G
E 129	Allura Red AC
E 131	Patent Blue V
E 132	Indigotine, indigo carmine
E 133	Brilliant Blue FCF
E 140	Chlorophylls and chlorophyllins:
	(i) Chlorophylls
	(ii) Chlorophyllins
E 141	Copper complexes of chlorophylls and chlorophyllins:
	(i) Copper complexes of chlorophylls
	(ii) Copper complexes of chlorophyllins
E 142	Green S
E 150a	Plain caramel
E 150b	Caustic sulphite caramel
E 150c	Ammonia caramel
E 150d	Sulphite ammonia caramel
E 151	Brilliant Black BN, Black PN
E 153	Vegetable carbon
E 154	Brown FK

E 155	Brown HT
E 160a	Carotenes:
	(i) Mixed carotenes
	(ii) Beta-carotene
E 160b	Annatto, bixin, norbixin
E 160c	Paprika extract, capsanthin, capsorubin
E 160d	Lycopene
E 160e	Beta-apo-8'-carotenal (C30)
E 160f	Ethyl ester of beta-apo-8'-carotenic acid (C30)
E 161b	Lutein
E 161g	Canthaxanthin
E 162	Beetroot Red, betanin
E 163	Anthocyanins
E 170	Calcium carbonates:
	(i) Calcium carbonate
	(ii) Calcium hydrogen carbonate
E 171	Titanium dioxide
E 172	Iron oxides and hydroxides
E 173	Aluminium
E 174	Silver
E 175	Gold
E 180	Litholrubine BK
E 200	Sorbic acid
E 202	Potassium sorbate
E 203	Calcium sorbate
E 210	Benzoic acid
E 211	Sodium benzoate
E 212	Potassium benzoate
E 213	Calcium benzoate
E 214	Ethyl p-hydroxybenzoate
E 215	Sodium ethyl p-hydroxybenzoate
E 216	Propyl p-hydroxybenzoate
E 217	Sodium propyl p-hydroxybenzoate
E 218	Methyl p-hydroxybenzoate
E 219	Sodium methyl p-hydroxybenzoate
E 220	Sulphur dioxide
E 221	Sodium sulphite
E 222	Sodium hydrogen sulphite
E 223	Sodium metabisulphite
E 224	Potassium metabisulphite
E 226	Calcium sulphite
E 227	Calcium hydrogen sulphite
E 228	Potassium hydrogen sulphite
E 230	Biphenyl, diphenyl
E 231	Orthophenyl phenol

E 232	Sodium orthophenyl phenol
E 234	Nisin
E 235	Natamycin
E 239	Hexamethylene tetramine
E 242	Dimethyl dicarbonate
E 249	Potassium nitrite
E 250	Sodium nitrite
E 251	Sodium nitrate
E 252	Potassium nitrate
E 260	Acetic acid
E 261	Potassium acetate
E 262	Sodium acetates:
	(i) Sodium acetate
	(ii) Sodium hydrogen acetate (sodium diacetate)
E 263	Calcium acetate
E 270	Lactic acid
E 280	Propionic acid
E 281	Sodium propionate
E 282	Calcium propionate
E 283	Potassium propionate
E 284	Boric acid
E 285	Sodium tetraborate (borax)
E 290	Carbon dioxide
E 296	Malic acid
E 297	Fumaric acid
E 300	Ascorbic acid
E 301	Sodium ascorbate
E 302	Calcium ascorbate
E 304	Fatty acid esters of ascorbic acid:
	(i) Ascorbyl palmitate
	(ii) Ascorbyl stearate
E 306	Tocopherol rich extract
E 307	Alpha-tocopherol
E 308	Gamma-tocopherol
E 309	Delta-tocopherol
E 310	Propyl gallate
E 311	Octyl gallate
E 312	Dodecyl gallate
E 315	Erythorbic acid
E 316	Sodium erythorbate
E 320	Butylated hydroxyanisole (BHA)
E 321	Butylated hydroxytoluene (BHT)
E 322	Lecithins
E 325	Sodium lactate
E 326	Potassium lactate

E 327	Calcium lactate
E 330	Citric acid
E 331	Sodium citrates: (i) Monosodium citrate (ii) Disodium citrate (iii) Trisodium citrate
E 332	Potassium citrates: (i) Monopotassium citrate (ii) Tripotassium citrate
E 333	Calcium citrates: (i) Monocalcium citrate (ii) Dicalcium citrate (iii) Tricalcium citrate
E 334	Tartaric acid (L(+)-)
E 335	Sodium tartrates: (i) Monosodium tartrate (ii) Disodium tartrate
E 336	Potassium tartrates: (i) Monopotassium tartrate (ii) Dipotassium tartrate
E 337	Sodium potassium tartrate
E 338	Phosphoric acid
E 339	Sodium phosphates: (i) Monosodium phosphate (ii) Disodium phosphate (iii) Trisodium phosphate
E 340	Potassium phosphates: (i) Monopotassium phosphate (ii) Dipotassium phosphate (iii) Tripotassium phosphate
E 341	Calcium phosphates: (i) Monocalcium phosphate (ii) Dicalcium phosphate (iii) Tricalcium phosphate
E 343	Magnesium phosphates (i) Monomagnesium phosphate (ii) Dimagnesium phosphate
E 350	Sodium malates: (i) Sodium malate (ii) Sodium hydrogen malate
E 351	Potassium malate
E 352	Calcium malates: (i) Calcium malate (ii) Calcium hydrogen malate
E 353	Metatartaric acid

E 354	Calcium tartrate
E 355	Adipic acid
E 356	Sodium adipate
E 357	Potassium adipate
E 363	Succinic acid
E 380	Triammonium citrate
E 385	Calcium disodium ethylene diamine tetra-acetate (Calcium disodium EDTA)
E 400	Alginic acid
E 401	Sodium alginate
E 402	Potassium alginate
E 403	Ammonium alginate
E 404	Calcium alginate
E 405	Propane-1,2-diol alginate
E 406	Agar
E 407	Carrageenan
E 407a	Processed eucheuma seaweed
E 410	Locust bean gum
E 412	Guar gum
E 413	Tragacanth
E 414	Acacia gum (gum arabic)
E 415	Xanthan gum
E 416	Karaya gum
E 417	Tara gum
E 418	Gellan gum
E 420	Sorbitol: (i) Sorbitol (ii) Sorbitol syrup
E 421	Mannitol
E 422	Glycerol
E 425	Konjac: (i) Konjac gum (ii) Konjac glucomannane
E 431	Polyoxyethylene (40) stearate
E 432	Polyoxyethylene sorbitan monolaurate (polysorbate 20)
E 433	Polyoxyethylene sorbitan monooleate (polysorbate 80)
E 434	Polyoxyethylene sorbitan monopalmitate (polysorbate 40)
E 435	Polyoxyethylene sorbitan monostearate (polysorbate 60)
E 436	Polyoxyethylene sorbitan tristearate (polysorbate 65)
E 440	Pectins: (i) Pectin (ii) Amidated pectin
E 442	Ammonium phosphatides
E 444	Sucrose acetate isobutyrate
E 445	Glycerol esters of wood rosins

E 450	Diphosphates: (i) Disodium diphosphate (ii) Trisodium diphosphate (iii) Tetrasodium diphosphate (iv) Tetrapotassium diphosphate (v) Dicalcium diphosphate (vi) Calcium dihydrogen diphosphate
E 451	Triphosphates: (i) Pentasodium triphosphate (ii) Pentapotassium triphosphate
E 452	Polyphosphates: (i) Sodium polyphosphate (ii) Potassium polyphosphate (iii) Sodium calcium polyphosphate (iv) Calcium polyphosphate
E 459	Beta-cyclodextrine
E 460	Cellulose: (i) Microcrystalline cellulose (ii) Powdered cellulose
E 461	Methyl cellulose
E 463	Hydroxypropyl cellulose
E 464	Hydroxypropyl methyl cellulose
E 465	Ethyl methyl cellulose
E 466	Carboxy methyl cellulose Sodium carboxy methyl cellulose
E 468	Cross linked sodium carboxy methyl cellulose
E 469	Enzymatically hydrolysed carboxy methyl cellulose
E 470a	Sodium, potassium and calcium salts of fatty acids
E 470b	Magnesium salts of fatty acids
E 471	Mono- and diglycerides of fatty acids
E 472a	Acetic acid esters of mono- and diglycerides of fatty acids
E 472b	Lactic acid esters of mono- and diglycerides of fatty acids
E 472c	Citric acid esters of mono- and diglycerides of fatty acids
E 472d	Tartaric acid esters of mono- and diglycerides of fatty acids
E 472e	Mono- and diacetyl tartaric acid esters of mono- and diglycerides of fatty acids
E 472f	Mixed acetic and tartaric acid esters of mono- and diglycerides of fatty acids
E 473	Sucrose esters of fatty acids
E 474	Sucroglycerides
E 475	Polyglycerol esters of fatty acids
E 476	Polyglycerol polyricinoleate
E 477	Propane-1,2-diol esters of fatty acids
E 479b	Thermally oxidised soya bean oil interacted with mono- and diglycerides of fatty acids

E 481	Sodium stearoyl-2-lactylate
E 482	Calcium stearoyl-2-lactylate
E 483	Stearyl tartrate
E 491	Sorbitan monostearate
E 492	Sorbitan tristearate
E 493	Sorbitan monolaurate
E 494	Sorbitan monooleate
E 495	Sorbitan monopalmitate
E 500	Sodium carbonates:
	(i) Sodium carbonate
	(ii) Sodium hydrogen carbonate
	(iii) Sodium sesquicarbonate
E 501	Potassium carbonates:
	(i) Potassium carbonate
	(ii) Potassium hydrogen carbonate
E 503	Ammonium carbonates:
	(i) Ammonium carbonate
	(ii) Ammonium hydrogen carbonate
E 504	Magnesium carbonates:
	(i) Magnesium carbonate
	(ii) Magnesium hydroxide carbonate
E 507	Hydrochloric acid
E 508	Potassium chloride
E 509	Calcium chloride
E 511	Magnesium chloride
E 512	Stannous chloride
E 513	Sulphuric acid
E 514	Sodium sulphates:
	(i) Sodium sulphate
	(ii) Sodium hydrogen sulphate
E 515	Potassium sulphates:
	(i) Potassium sulphate
	(ii) Potassium hydrogen sulphate
E 516	Calcium sulphate
E 517	Ammonium sulphate
E 521	Aluminium sodium sulphate
E 522	Aluminium potassium sulphate
E 523	Aluminium ammonium sulphate
E 524	Sodium hydroxide
E 525	Potassium hydroxide
E 526	Calcium hydroxide
E 527	Ammonium hydroxide
E 528	Magnesium hydroxide
E 529	Calcium oxide
E 530	Magnesium oxide

E 535	Sodium ferrocyanide
E 536	Potassium ferrocyanide
E 538	Calcium ferrocyanide
E 541	Sodium aluminium phosphate, acidic
E 551	Silicon dioxide
E 552	Calcium silicate
E 553a	(i) Magnesium silicate (ii) Magnesium trisilicate
E 553b	Talc
E 554	Sodium aluminium silicate
E 555	Potassium aluminium silicate
E 556	Calcium aluminium silicate
E 558	Bentonite
E 559	Aluminium silicate (Kaolin)
E 570	Fatty acids
E 574	Gluconic acid
E 575	Glucono-delta-lactone
E 576	Sodium gluconate
E 577	Potassium gluconate
E 578	Calcium gluconate
E 579	Ferrous gluconate
E 585	Ferrous lactate
E 620	Glutamic acid
E 621	Monosodium glutamate
E 622	Monopotassium glutamate
E 623	Calcium diglutamate
E 624	Monoammonium glutamate
E 625	Magnesium diglutamate
E 626	Guanylic acid
E 627	Disodium guanylate
E 628	Dipotassium guanylate
E 629	Calcium guanylate
E 630	Inosinic acid
E 631	Disodium inosinate
E 632	Dipotassium inosinate
E 633	Calcium inosinate
E 634	Calcium 5'-ribonucleotides
E 635	Disodium 5'-ribonucleotides
E 640	Glycine and its sodium salt
E 650	Zinc acetate
E 900	Dimethyl polysiloxane
E 901	Beeswax, white and yellow
E 902	Candelilla wax
E 903	Carnauba wax
E 904	Shellac

E 912	Montan acid esters
E 914	Oxidised polyethylene wax
E 920	L-cysteine
E 927b	Carbamide
E 938	Argon
E 939	Helium
E 941	Nitrogen
E 942	Nitrous oxide
E 943a	Butane
E 943b	Iso-butane
E 944	Propane
E 948	Oxygen
E 949	Hydrogen
E 950	Acesulfame K
E 951	Aspartame
E 952	Cyclamic acid and its Na and Ca salts
E 953	Isomalt
E 954	Saccharin and its Na, K and Ca salts
E 957	Thaumatin
E 959	Neohesperidine DC
E 965	Maltitol:
	(i) Maltitol
	(ii) Maltitol syrup
E 966	Lactitol
E 967	Xylitol
E 999	Quillaia extract
E 1103	Invertase
E 1105	Lysozyme
E 1200	Polydextrose
E 1201	Polyvinylpyrrolidone
E 1202	Polyvinylpolypyrrolidone
E 1404	Oxidised starch
E 1410	Monostarch phosphate
E 1412	Distarch phosphate
E 1413	Phosphated distarch phosphate
E 1414	Acetylated distarch phosphate
E 1420	Acetylated starch
E 1422	Acetylated distarch adipate
E 1440	Hydroxy propyl starch
E 1442	Hydroxy propyl distarch phosphate
E 1450	Starch sodium octenyl succinate
E 1451	Acetylated oxidised starch
E 1505	Triethyl citrate
E 1518	Glyceryl triacetate (triacetin)
E 1520	Propane-1, 2-diol (propylene glycol)

5

Contaminants

S. A. Slorach, National Food Administration, Uppsala

5.1 Introduction

This chapter briefly describes how European Community (EC) legislation on chemical contaminants in foodstuffs is developed, adopted and enforced. In this context the term 'chemical contaminants' covers residues of pesticides and veterinary drugs, heavy metals, mycotoxins and nitrate.

Community procedures for preparing and adopting legislation on contaminants in foodstuffs involve, among others, the European Commission ('the Commission') and its scientific advisory bodies, the Council of the European Union (the 'Council'), the fifteen member states of the European Union (EU) and the European Parliament. Proposals for new legislation are made by the Commission. Decisions on new legislation on contaminants in foodstuffs are generally made by the Council, unless authority to do so has been delegated to the Commission. In addition to the complications arising from the above situation, different procedures are used for different classes of substance (for example, the procedure for elaborating legislation on mycotoxins is different from that for veterinary drug residues) and the procedures have changed over the years and are still changing. Responsibility for initiating work on legislation on pesticide residues, mycotoxins, heavy metals and nitrate rests with the Commission's Directorate-General for Health and Consumer Protection (usually referred to as DG SANCO), whereas veterinary drug residues are the responsibility of the Commission's Directorate-General for Enterprise. A brief outline of the procedures used in developing and enforcing regulations on chemical contaminants is given here and readers wishing to immerse themselves in the detail, of which there is plenty, should consult the references given at the end of the chapter.

All Community legislation and proposals for legislation are published in the *Official Journal of the European Communities* (the *Official Journal*). Legislative instruments are mainly of two types – Regulations and Directives. Regulations are binding in their entirety and apply directly (verbatim) in all member states. Directives are binding, as to the result to be achieved, but allow the national authorities the choice of form and methods: they are transposed into the national legislation of the member states and may therefore differ slightly from country to country. In addition, there are Decisions, which are binding in their entirety upon those to whom they are addressed. Recommendations and Opinions, on the other hand, have no binding force.

5.2 Scientific advisory committees

Most of the expert advice that forms the scientific basis of EC legislation on chemical contaminants in food is provided by the Commission's scientific advisory committees working in the food safety area. The experts on these committees are expected to provide independent advice and not represent their countries or the organisations that employ them. The results of the deliberations of the scientific advisory committees are published and are also available on the Internet at DG SANCO's website (http://europa.eu.int/comm/dgs/health_consumer/index_en.htm).

In 1997, the Commission transferred responsibility for these committees to DG SANCO and restructured the system. It established a Scientific Steering Committee and defined the mandates for eight scientific advisory committees (Commission Decision 97/579) (see Fig. 5.1). The aim of the reform of the system was to increase the independence of the expert committees from national government and other interests and to increase transparency. Members of the expert committees are required to declare any interests they may have that could affect their impartiality in dealing with a particular subject or substance on a committee's agenda.

The committees most involved in setting limits for contaminants in food are the Scientific Committee on Food, the Scientific Committee on Veterinary Measures Relating to Public Health and the Scientific Committee on Plants. Prior to autumn 1997, there was a separate Scientific Committee for Pesticides, but its responsibilities are now included in the mandate of the Scientific Committee on Plants. In order to facilitate their work, many of the scientific committees have subsidiary working groups which carry out a lot of the preparatory work that needs to be done (preparation of working documents, etc.) before opinions can be given by the committee. The Scientific Committee on Food has a Working Group on Contaminants (see Fig. 5.2).

In November 2000, the Commission proposed the creation of a European Food Authority (EFA), which would take over responsibility for much of the food safety risk assessment work at present carried out by the Commission and

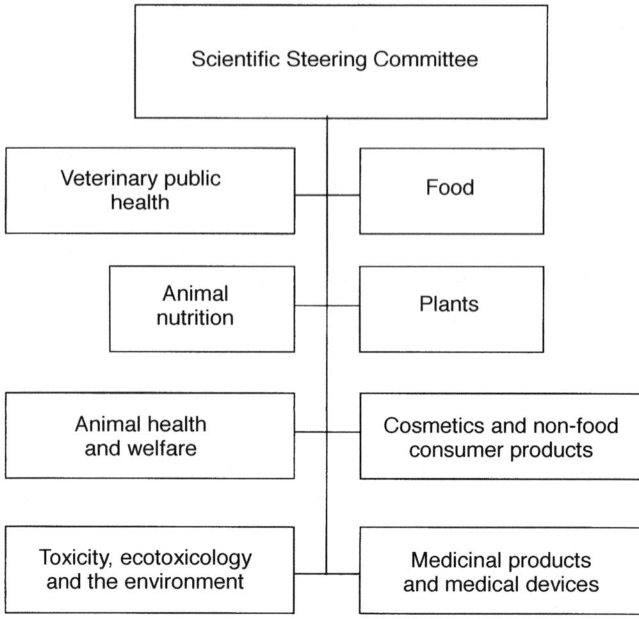

Fig. 5.1 European Commission scientific advisory committees.

its advisory bodies. Thus if and when the EFA is established and comes into operation (perhaps in 2002), it will assume responsibility for providing risk assessments as a basis for EC legislation on contaminants in food.

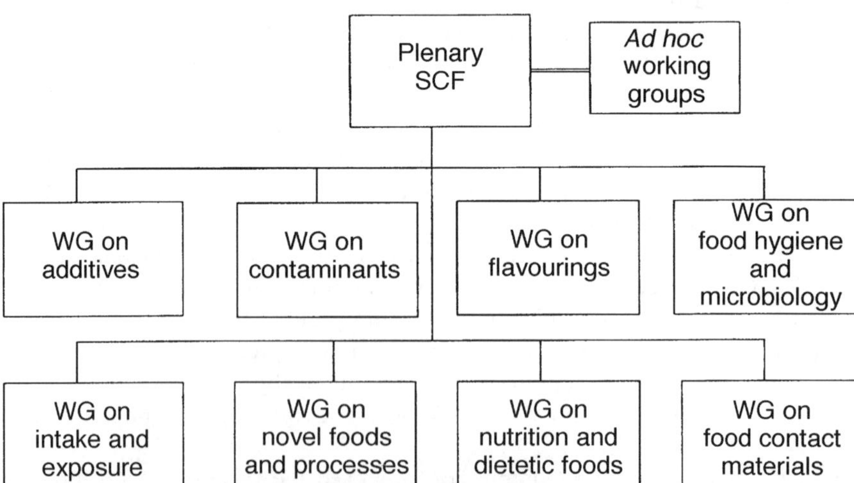

Fig. 5.2 SCF Working Group structure.

5.3 Pesticide residues

The procedure used to set permanent maximum residue levels (MRLs) for pesticide residues is described briefly below. In parallel with this, proposals for provisional MRLs are prepared in connection with the work on regulating the placing of plant protection products on the market according to Council Directive 91/414.

5.3.1 Toxicological evaluations

The DG SANCO's Working Group on Pesticide Residues proposes MRLs for pesticide residues in foodstuffs. For each pesticide, MRLs are proposed for residues in the relevant individual commodities/crops. Since a large number of pesticides have to be dealt with, the workload is spread by appointing each member state in the working group as rapporteur for a number of specified pesticides.

When proposing an MRL, the rapporteur member state (RMS) identifies the Acceptable Daily Intake (ADI) and Acute Reference Dose (ARFD) for humans that is valid for the pesticide in question. The ADI thus identified is often the same as that recommended by the Joint FAO/WHO Meetings on Pesticide Residues (JMPR), whose recommendations on ADIs and MRLs are used within the Codex Alimentarius system. If the ADI proposed by the RMS is not that recommended by JMPR, the RMS has to provide an explanation for the difference. The other member states comment on the RMS proposal at meetings of the Working Group on Pesticide Residues. If the member states cannot reach agreement on the evaluation, the matter is referred to one or more of the Commission's scientific advisory committees. Prior to autumn 1997, such questions were referred to the Scientific Committee for Pesticides, but since then they have been referred to the Scientific Committee on Plants.

5.3.2 Residue data

Data on the levels of residues found in supervised trials, in which the pesticide is used according to Good Agricultural Practice (GAP), are the main basis for proposing MRLs. Such trials must include studies that reflect the use that results in the highest residue levels or critical GAP. MRLs are established for individual crops or groups of crops. In certain cases, data from supervised trials on one crop may be extrapolated to another crop. The results of the supervised trials are evaluated by the RMS and then discussed by the Working Group on Pesticide Residues.

5.3.3 Pesticide intake calculations

Although the main basis for proposing MRLs is the data from supervised trials, when considering the proposals member states calculate the theoretical maximum daily intake that could occur if the proposed MRLs were adopted, using the World Health Organization (WHO) European Diet or national dietary information as the basis for their calculations. If the theoretical maximum daily intake exceeds the proposed ADI, member states may not be prepared to accept the proposed MRL without further refinement of the intake calculations. The EC takes into consideration not only chronic exposure to pesticide residues via foodstuffs, but also acute exposure. The acute exposure is assessed in accordance with the procedures and practices used in the EC, taking account of guidelines published by the WHO.

5.3.4 Preparation and adoption of legislation

Nowadays, EC MRLs are laid down according to the procedures in Directive 97/41, which amended Directives 76/895, 86/362, 86/363 and 90/642. Earlier decisions on MRLs were made by a Council procedure, i.e. the Council made the final decision. However, they are now usually decided by the Commission according to a regulatory committee decision procedure (Procedure IIIb) involving the Commission's Standing Committee on Plant Health. This Standing Committee consists of representatives of the member states, but is chaired by the Commission. If the members of this standing committee cannot agree, i.e. there is not a qualified majority for a proposal, the question is referred to the Council for a decision. The Council shall act by a qualified majority. If the Council fails to reply within the stipulated time limit, the proposal shall be adopted by the Commission, except where the Council has decided against the measure by a simple majority. Before the proposals for MRLs are notified to the World Trade Organization, they are sent to the Scientific Committee on Plants.

5.3.5 EC legislation on maximum residue levels

Council Directive 76/895 (and Directives 81/36, 82/528, 88/298, 2000/24, 2000/57 and 2000/82, which contain amendments to that directive) contains recommendations for MRLs for pesticide residues in or on fruit and vegetables. However, these MRLs are not mandatory and member states may set higher MRLs in their national legislation, but not lower levels.

Council Directives 86/362, 86/363 and 90/642 contain MRLs for pesticide residues in or on cereals, foods of animal origin and fruit and vegetables, respectively. These MRLs are mandatory and must be incorporated into the national legislation of the member states. These directives have been amended several times (Directives 88/298, 93/57, 94/29, 95/39, 96/33, 98/82, 1999/71 2000/24, 2000/42, 2000/48, 2000/58, 2000/81 and 2000/82 concerning MRLs for cereals and foods of animal origin and Directives 93/58, 94/30, 95/38, 95/61, 96/32, 98/82, 1999/71, 2000/24, 2000/42, 2000/48, 2000/57, 2000/58, 2000/81 and 2000/82 concerning MRLs for fruit and vegetables). Unfortunately, official consolidated versions of the amended directives are not produced, which makes it difficult to get a complete picture of all the MRLs that have been adopted to date. However, an unofficial list of EC MRLs can be found on the website of DG SANCO.

5.3.6 Interaction with Codex

Many EU member states participate actively in the work of the Codex Committee on Pesticide Residues (CCPR), which is hosted by the Netherlands. Before and during each CCPR meeting, EC positions are coordinated as far as possible. In view of the special status attached to Codex MRLs since the signing of the Agreement on Sanitary and Phytosanitary Measures (the SPS Agreement), the EC now attaches great importance to Codex work.

5.3.7 Enforcement

Enforcement of EC regulations on pesticide residues in foodstuffs is the responsibility of the competent authorities in the member states. The authorities in each member state should ensure that the products produced in that country and those imported directly from countries outside the EU ('Third Countries') comply with EC legislation. In principle, when such a system is fully developed and operational in all member states, there should be no real need for countries to examine products coming from other member states.

Each year the Commission issues recommendations concerning coordinated Community monitoring programmes to ensure compliance with the MRLs in and on cereals, fruit and vegetables, etc. The latest recommendation was published as Commission Recommendation 2001/42 in January 2001. It contained detailed information on the pesticide residue/product combinations to be monitored, the number of samples to be taken by each country and quality control procedures for the analysis, etc. In addition to the recommended (minimum) monitoring programme, the different member states carry out their own individual programmes, which can vary in both content and scope. Member states are encouraged to publish the results of their pesticide control work and the results of the recommended Community programme are collected and published by the Commission.

The Commission's Food and Veterinary Office (FVO) carries out inspections in member states to check that they have incorporated EC legislation into their national regulations and that they are enforcing it properly. The reports of these inspections are placed on the Internet at DG SANCO's website.

5.4 Veterinary drug residues

5.4.1 Introduction

The procedures used within the Community for establishing MRLs for veterinary medicinal products in foodstuffs of animal origin are laid down in Council Regulation 2377/90, that came into force on 1 January 1992. The most important principle laid down in these procedures is that foodstuffs obtained from treated animals must not contain residues which might constitute a health hazard for the consumer. When calculating MRLs, the aim is to ensure that the total intake of residues of the substance via foodstuffs of animal origin does not exceed the ADI. For the purposes of these calculations, the body-weight of the consumer has been assumed to be 60 kg and the daily intakes of various foods has been assigned certain values, e.g. milk 1,500 g, muscle 300 g, liver 100 g. In cases where a substance is also used as a pesticide, the intake from such use is also taken into account. The procedures described below apply to veterinary drugs used as medicines, but not to the medical substances used as feed additives.

MRLs are determined by the Committee for Veterinary Medicinal Products (CVMP), and its Safety Working Party, attached to the European Medicine Evaluation Agency (EMEA) in London. Different procedures are used, depending on when the pharmacologically active substance was first authorised for use. For medicines containing substances authorised after 1 January 1992 ('new substances'), MRLs must be set at the European level for all pharmacologically active substances, including excipients, before approval procedures can be started in the member states. MRLs for medicines authorised before 1 January 1992 ('old substances') must have been evaluated by 31 December 1999 if their use after that date is to continue. Regardless of the procedure to be followed in setting the MRL, the manufacturer of a veterinary medicinal product must provide safety and residue dossiers containing the information required to set an ADI and MRLs. The safety dossier contains the pharmacodynamic, kinetic, metabolic and toxicity data and the residue dossier the data on kinetics, metabolism and residues, as well as the analytical method(s) for the substance.

Regulation 2377/90 contains the following four annexes in which the substances are listed after evaluation: I. Substances for which final MRLs have been fixed; II. Substances for which MRLs are not deemed necessary in order to protect public health; III. Substances with provisional MRLs – if a dossier is incomplete, the manufacturer may be given a set time (up to five years) in which to provide the necessary information; IV. Substances for which it is not possible, due to safety concerns, to set an MRL – the administration of substances listed in this annex is prohibited throughout the EU and the marketing authorisation for the medicines concerned has been withdrawn.

5.4.2 MRLs for substances authorised after 1 January 1992

Applications from industry are submitted to the EMEA. Since a large number of substances have to be dealt with, the workload is spread by appointing each member state represented in the CVMP as rapporteur or co-rapporteur for a number of specified veterinary drugs. Using the information in the dossier provided by the manufacturer, the (RMS) proposes an ADI for the drug in question and MRLs for relevant foodstuffs of animal origin (e.g. muscle, liver, milk), using the guidelines in Regulation 2377/90 and in the Rules Governing Medicinal Products in the European Community, Vol. VI (soon Vol. VIII). The ADI and MRLs are often, but not always, the same as those proposed by JECFA or JMPR for the same drug. In the CVMP other member states then comment on the ADI and MRLs proposed by the RMS. When the CVMP has reached agreement on MRLs for a drug, they are submitted to the Commission, for adoption by the Committee for the Adaptation to Technical Progress of the Directives on the Removal of Technical Barriers to Trade in the Veterinary Medical Products Sector. If a qualified majority of the member states in that committee supports adoption, the MRLs are then incorporated into the relevant annex to Council Regulation 2377/90. If a qualified majority is not obtained, the Commission proposes to the Council the measures to be adopted: the Council acts by a qualified majority. If the Council has not acted within three months, the proposed measures are adopted by the Commission, unless the Council has voted against them by a simple majority. All amendments to the annexes of Regulation 2377/90 are published in the *Official Journal*.

5.4.3 MRLs for substances authorised before 1 January 1992

After examining dossiers supplied by industry, RMSs in the CVMP's Safety Working Party (SWPV) propose ADIs and MRLs for substances authorised before 1 January 1992. These proposals are then discussed by the CVMP. When the CVMP has reached agreement on MRLs, they are then adopted by the procedure described above for 'new substances'.

5.4.4 Withdrawal periods

The 'withdrawal period' is the time between the last dose given to the animal and the time when the level of residues in the tissues (muscle, liver, kidney, skin/fat) or products (milk, eggs, honey) is lower than or equal to the MRL. For veterinary medicinal products intended to be marketed in only one member state, withdrawal periods are set at the national (member state) level. This is also the case for all the old substances authorised before 1 January 1992. For products intended to be used in more than one member state, the mutual recognition procedure has been obligatory since 1 January 1998. In this procedure evaluation is carried out in one country and the proposed withdrawal period is then accepted (or rejected) by other member states. If the proposal is not accepted, the matter can go to arbitration at EMEA. For products intended to be marketed throughout the whole of the European Union, withdrawal periods are determined by the CVMP by a central procedure analogous to that used for MRLs. This procedure must always be used for certain special groups of substances, e.g. those produced by biotechnology and innovative products.

5.4.5 Interaction with Codex

Many EU member states and the Commission take an active part in the Codex work on MRLs for veterinary drugs, especially in the Codex Committee on Residues of Veterinary Drugs in Food (CCRVDF). Before and during each CCRVDF meeting, EC positions are coordinated as far as possible. In view of the special status attached to Codex MRLs since the signing of the SPS Agreement, the EC now attaches great importance to Codex work. The MRLs proposed by JECFA and JMPR and discussed in CCRVDF are in many, but not all, cases accepted by the member states of the EU. The most notable exceptions to this in recent years are the MRLs for hormones used for growth promotion, which have been the subject of much acrimonious debate in Codex and also the subject of an SPS Dispute Panel.

5.4.6 Enforcement

Enforcement of EC regulations on veterinary drug residues in foodstuffs is the responsibility of the competent authorities in the member states. There are also regulations on checks to be carried out on live animals to check the absence of growth promoters whose use is prohibited in the EU. The authorities in each member state are required to ensure that the relevant products comply with EC legislation. Council Directive 96/23 and Commission Decision 97/747 lay down the measures to be taken to monitor residues of veterinary drugs in foods of animal origin and specify the minimum level and frequency of sampling for such control. Each year member states are required to submit their monitoring programmes to the Commission for approval and also to report the results of their monitoring work. The Commission's Food and Veterinary Office, based in Ireland, also carries out inspections in member states to ensure that EC legislation on veterinary drug residues is enforced.

5.5 Mercury and histamine in fishery products

Council Directive 91/493 lays down the health conditions for the production and placing on the market of fishery products. Chapter V of that directive contains, among other things, maximum limits for histamine in certain fish species and instructions for checking that these limits are not exceeded. That chapter also makes provision for the establishment of limits for the presence in fish of contaminants from the aquatic environment.

Commission Decision 93/351 lays down maximum limits for mercury in fishery products and methods of sampling and analysis to check compliance with these limits. This decision was made after consulting the Standing Veterinary Committee. The mean total mercury content of the edible parts of fishery products must not exceed 0.5 mg/kg of fresh weight. However, this average limit is increased to 1 mg/kg fresh weight for the edible parts of certain species listed in the annex to the decision (a revision of this list of fish species is under way). The higher limit applies to *inter alia* sharks, tuna, swordfish, halibut and pike. In future, the maximum levels for mercury in fish will be regulated in a similar way to that described below for other heavy metals.

5.6 Other chemical contaminants

5.6.1 General procedure

Council Regulation 315/93 lays down Community procedures for establishing maximum limits for contaminants (other than pesticide and veterinary drug residues) in food. The Scientific Committee on Food must be consulted on all questions that may have an effect on public health and this committee carries out the toxicological evaluations that underpin the limits set for contaminants. The scientific data that form the basis of the evaluations are obtained mainly from the scientific literature and from the member states. Data on human exposure to contaminants, such as nitrates, cadmium, aflatoxins and ochratoxin A, have been collected and collated in projects in the programme on scientific cooperation between the member states (known as SCOOP).

Proposals for new limits prepared by Commission working parties are submitted to the Standing Committee for Foodstuffs, which consists of representatives of the member states, but is chaired by the Commission. Decisions on new limits are usually made by the Commission according to a regulatory committee procedure (Procedure IIIb) – for details see section 5.3.4. The Commission publishes the limits as a regulation in the *Official Journal*. Methods of sampling and analysis to check compliance with the maximum levels laid down are also published.

5.6.2 Mycotoxins

Aflatoxins, ochratoxin A, patulin, nivalenol, deoxynivalenol, fumonisins and zearalenone have been evaluated by the Scientific Committee on Food. The question of maximum levels for some of these mycotoxins in foodstuffs has been discussed for several years in the Committee of Experts – Working Party on Agricultural Contaminants under DG VI (now under DG SANCO). Proposals from this committee are then considered by the Standing Committee on Foodstuffs, prior to adoption by the Commission as Commission regulations.

Maximum levels for aflatoxin MI in milk and for aflatoxin B1 and the sum of aflatoxins B1, B2, G1 and G2 in groundnuts and certain other foods were laid down in Commission Regulation 1525/98 (which amended Regulation 194/97) and came into force on 1 January 1999. The Commission is expected to adopt maximum levels for aflatoxins in spices in the near future.

The question of maximum levels for ochratoxin A and patulin in certain foods has been under discussion for some time and a decision is expected soon.

A recommendation has recently been made regarding the maximum level of deoxynivalenol in cereal products.

5.6.3 Heavy metals other than mercury

Discussions on limits for lead and cadmium in a wide range of foodstuffs have been going on for several years in a working party under DG III (now under DG SANCO). The Scientific Committee on Food has carried out toxicological evaluations on these metals. As yet, no limits for these metals have been adopted by the Commission, but a decision is expected soon.

5.6.4 Nitrate in lettuce and spinach

Proposals for limits for nitrate in certain vegetables were prepared by a Committee of Experts in the Working Party on Agricultural Contaminants under the former Directorate-General VI (now under DG SANCO). The proposals were then considered under the above-mentioned procedure and the Commission has issued Regulation 194/97 setting maximum levels for nitrates in lettuce and spinach.

5.6.5 Interaction with Codex

Within the Codex system, the contaminants considered in this section are mainly dealt with by the Codex Committee on Food Additives and Contaminants (CCFAC), which is hosted by the Netherlands. Many of the member states of the EU are very active in CCFAC. For example, Denmark and the Netherlands have been instrumental in developing the Codex General Standard on Contaminants and Toxins and draft limits for lead in various foods. Sweden has developed a proposal for a limit for ochratoxin A in cereals and cereal products and France has proposed a maximum level for patulin in apple juice.

5.7 Future trends

The EC has established procedures for preparing, adopting and enforcing legislation on limits for various chemical contaminants in foodstuffs. In recent years, the work of the scientific advisory committees, which provide the scientific basis for most of the limits, has become more independent and transparent and this trend is likely to continue.

Much work still remains to be done on limits for pesticide residues. In addition to new substances, there is an urgent need to re-evaluate many of the older pesticides in the light of new toxicological data. Council Directive 91/414 concerning the placing of plant protection products on the market provides for the Commission to assess the safety aspects of pesticides. The programme of work for setting MRLs for pesticide residues is gradually being aligned with that on the evaluation of pesticides according to Directive 91/414. A timetable for the work planned for the next few years has been agreed. Continued cooperation with countries outside the EU should expedite matters.

The setting of MRLs for residues of veterinary drugs has been simplified somewhat since 1999, when the evaluation of 'old substances' was completed and all MRLs are now developed and adopted by a unified central procedure.

Much work has still to be done on the preparation and adoption of maximum levels for mycotoxins, heavy metals and other contaminants, such as PCBs and dioxins. Here one of the main factors delaying progress is the lack of data for toxicological evaluations and setting tolerable daily or weekly intakes. Furthermore, there is a lack of reliable data on levels of contaminants in individual foodstuffs and on dietary intakes of such substances.

Many EU member states already play an active role in the development of Codex limits for contaminants. It is foreseen that this will continue and that coordination between the member states and the European Commission on Codex matters will further improve. Limits for many substances mentioned above, e.g. ochratoxin A, lead, cadmium and some pesticides, are being discussed in parallel in the EC and in Codex and often by the same people. This facilitates coordination of the work in these different fora and should hopefully expedite the establishment of limits that can be widely accepted.

5.8 References

All the references to legislation published in the *Official Journal of the European Communities* are listed here in chronological order.

Council Directive 76/895/EEC of 23 November 1976 relating to the fixing of maximum levels for pesticide residues in or on fruit and vegetables, *Official Journal of the European Communities*, L 340, pp. 26–31.

Council Directive 81/36/EEC of 9 February 1981 amending annex II to Directive 76/895/EEC relating to the fixing of maximum levels for pesticide residues in and on fruit and vegetables, *Official Journal of the European Communities*, L 46, pp. 33–4.

Council Directive 82/528/EEC of 19 July 1982 amending annex II to Directive 76/895/EEC relating to the fixing of maximum levels for pesticide residues in and on fruit and vegetables, *Official Journal of the European Communities*, L 234, pp. 1–4.

Council Directive 86/362/EEC of 24 July 1986 on the fixing of maximum levels for pesticide residues in and on cereals, *Official Journal of the European Communities*, L 221, pp. 37–42.

Council Directive 86/363/EEC of 24 July 1986 on the fixing of maximum levels for pesticide residues in and on foods of animal origin, *Official Journal of the European Communities*, L 221, pp. 43–7.

Council Directive 88/298/EEC of 16 May 1988 amending annex II to Directives 76/895/EEC and 86/362 relating to the fixing of maximum levels for pesticide residues in and on fruit and vegetables and cereals, respectively, *Official Journal of the European Communities*, L 126, pp. 53–4.

Council Regulation 90/2377 EEC of 26 June 1990 laying down a Community procedure for the establishment of maximum residue limits of veterinary medicinal products in foodstuffs of animal origin, *Official Journal of the European Communities*, L 224, pp. 7–14.

Council Directive 90/642/EEC of 27 November 1990 on the fixing of maximum levels for pesticide residues in and on certain products of plant origin, including fruit and vegetables, *Official Journal of the European Communities*, L 350, pp. 71–9.

Council Directive 91/493/EEC of 22 July 1991 laying down the health conditions for the production and placing on the market of fishery products, *Official Journal of the European Communities*, L 268, pp. 15–34.

Council Directive 91/414/EEC of 15 July 1991 concerning the placing of plant protection products on the market, *Official Journal of the European Communities*, L 230, pp. 1–32.

Rules governing medicinal products in the European Community. Volume VI: Establishment by the European Community of maximum residue limits (MRLs) for residues of veterinary medicinal products in foodstuffs of animal origin, Commission of the European Communities, Directorate-General for Internal Market and Industrial Affairs, October 1991.

Council Regulation 93/315/EEC of 8 February 1993 laying down Community procedures for contaminants in food, *Official Journal of the European Communities*, L 37, pp. 1–3.

Commission Decision 93/351/EEC of 19 May 1993 determining analysis methods, sampling plans and maximum limits for mercury in fishery products, *Official Journal of the European Communities*, L 144, pp. 23–4.

Council Directive 93/57/EEC of 29 June 1993 amending the annexes to Directives 86/362/EEC and 86/363/EEC on the fixing of maximum levels for pesticide residues in and on cereals and foodstuffs of animal origin, respectively, *Official Journal of the European Communities*, L 211, pp. 1–5.

Council Directive 93/58/EEC of 29 June 1993 amending annex II to Directive 76/895/EEC relating to the fixing of maximum levels for pesticide residues in or on fruit and vegetables and the annex to Directive 90/462/EEC relating to the fixing of maximum levels for pesticide residues in and on certain products of plant origin, including fruit and vegetables, and providing for the establishment of a first list of maximum levels, *Official Journal of the European Communities*, L 211, pp. 6–39.

Council Directive 94/29/EC of 23 June 1994 amending the annexes to Directives 86/362/EEC and 86/363/EEC on the fixing of maximum levels for pesticide residues in and on cereals and foodstuffs of animal origin, respectively, *Official Journal of the European Communities*, L 189, pp. 67–9.

Council Directive 94/30/EC of 23 June 1994 amending annex II to Directive 90/642/EEC relating to the fixing of maximum levels for pesticide residues in and on certain products of plant origin, including fruit and vegetables, and providing for the establishment of a list of maximum levels, *Official Journal of the European Communities*, L 189, pp. 70–83.

Council Directive 95/38/EC of 17 July 1995 amending annexes I and II to Directive 90/642/EEC on the fixing of maximum levels for pesticide residues in and on certain products of plant origin, including fruit and vegetables, and providing for the establishment of a list of maximum levels, *Official Journal of the European Communities*, L 197, pp. 14–28.

Council Directive 95/39/EC of 17 July 1995 amending the annexes to Directives 86/362/EEC and 86/363/EEC on the fixing of maximum levels for pesticide residues in and on cereals and foodstuffs of animal origin, *Official Journal of the European Communities*, L 197, pp. 29–31.

Council Directive 95/61/EC of 29 November 1995 amending annex II to Directive 90/642/EEC on the fixing of maximum levels for pesticide residues in and on certain products of plant origin, including fruit and vegetables, *Official Journal of the European Communities*, L 292, pp. 27–30.

Council Directive 96/23/EC of 29 April 1996 on measures to monitor certain substances and residues thereof in live animals and animal products and repealing Directives 85/358/EEC and 86/469/EEC and Decisions 89/187/EEC and 91/664/EEC, *Official Journal of the European Communities*, L 125, pp. 10–32.

Council Directive 96/32/EC of 21 May 1996 amending annex II to Directive 76/895/EEC relating to fixing of maximum limits for pesticide residues in and on fruit and vegetables and annex II to Directive 90/642/EEC relating to the fixing of maximum levels for pesticide residues in and on certain products of plant origin, including fruit and vegetables, and providing for the establishment of a list of maximum levels, *Official Journal of the European Communities*, L 144, pp. 12–34.

Council Directive 96/33/EC of 21 May 1996 amending annexes to Directives 86/362/EEC and 86/363/EEC on the fixing of maximum levels for pesticide residues in and on cereals and foodstuffs of animal origin, *Official Journal of the European Communities*, L 144, pp. 35–8.

Commission Regulation 194/97/EC of 31 January 1997 setting maximum levels for certain contaminants in foodstuffs, *Official Journal of the European Communities*, L 31, pp. 48–50.

Council Directive 97/41/EC of 25 June 1997 amending Directives 76/895/EEC, 86/362/EEC, 86/363/EEC and 90/462/EEC relating to the fixing of maximum levels for pesticide residues in and on, respectively, fruit and vegetables, cereals, foodstuffs of animal origin, and certain products of plant origin, including fruit and vegetables, *Official Journal of the European Communities*, L 184, pp. 33–49.

Commission Decision 97/579/EC of 23 July 1997 setting up Scientific Committees in the field of consumer health and food safety, *Official Journal of the European Communities*, L 237, pp. 18–23.

Commission Decision 97/747/EC of 27 October 1997 fixing the levels and frequencies of sampling provided for by Council Directive 96/23/EC for the monitoring of certain substances and residues thereof in certain animal products, *Official Journal of the European Communities*, L 303, pp. 12–15.

Commission Regulation 1525/98/EC of 16 July 1998 amending Regulation (EC) No. 194/97 of 31 January 1997 setting maximum levels for certain contaminants in foodstuffs, *Official Journal of the European Communities*, L 201, pp. 43–6.

Commission Directive 98/82/EC of 27 October 1998 amending the annexes to Council Directives 86/362/EEC, 86/363/EEC and 90/642/EEC on the fixing of maximum levels for pesticide residues in and on cereals, foodstuffs of animal origin and certain products of plant origin, including fruit and vegetables, respectively, *Official Journal of the European Communities*, L 290, pp. 25–54.

Commission Directive 1999/71/EC of 14 July 1999 amending the annexes to Council Directives 86/362/EEC, 86/363/EEC and 90/642/EEC on the fixing of maximum levels for pesticide residues in and on cereals, foodstuffs of animal origin and certain products of plant origin, including fruit and vegetables, respectively, *Official Journal of the European Communities*, L 194, pp. 36–44.

Commission Directive 2000/24/EC of 28 April 2000 amending the annexes to Council Directives 76/895/EEC, 86/362/EEC, 86/363/EEC and 90/642/EEC on the fixing of maximum levels for pesticide residues in and on cereals, foodstuffs of animal origin and certain products of plant origin, including fruit and vegetables, respectively, *Official Journal of the European Communities*, L 107, pp. 28–37.

Commission Directive 2000/42/EC of 22 June 2000 amending the annexes to Council Directives 86/362/EEC, 86/363/EEC and 90/642/EEC on the fixing of maximum levels for pesticide residues in and on cereals, foodstuffs of animal origin and certain products of plant origin, including fruit and vegetables, respectively, *Official Journal of the European Communities*, L 158, pp. 51–75.

Commission Directive 2000/48/EC of 25 July 2000 amending the annexes to Council Directives 86/362/EEC and 90/642/EEC on the fixing of maximum levels for pesticide residues in and on cereals and certain products of plant origin, including fruit and vegetables, respectively, *Official Journal of the European Communities*, L 197, pp. 26–31.

Commission Directive 2000/57/EC of 22 September 2000 amending the annexes to Council Directives 76/895/EEC and 90/642/EEC on the fixing of maximum levels for pesticide residues in and on fruit and vegetables and certain products of plant origin, including fruit and vegetables, respectively, *Official Journal of the European Communities*, L 244, pp. 76–7.

Commission Directive 2000/58/EC of 22 September 2000 amending the annexes to Council Directives 86/362/EEC, 86/363/EEC and 90/642/EEC on the fixing of maximum levels for pesticide residues in and on cereals, foodstuffs of animal origin and certain products of plant origin, including fruit and vegetables, respectively, *Official Journal of the European Communities*, L 244, pp. 78–83.

Commission Directive 2000/81/EC of 18 December 2000 amending the annexes to Council Directives 86/362/EEC, 86/363/EEC and 90/642/EEC on the fixing of maximum levels for pesticide residues in and on cereals, foodstuffs of animal origin and certain products of plant origin, including fruit and vegetables, respectively, *Official Journal of the European Communities*, L 326, pp. 56–62.

Commission Directive 2000/82/EC of 20 December 2000 amending the annexes to Council Directives 76/895/EEC, 86/362/EEC, 86/363/EEC and 90/642/EEC on the fixing of maximum levels for pesticide residues in and on cereals, foodstuffs of animal origin and certain products of plant origin, including fruit and vegetables, respectively, *Official Journal of the European Communities*, L 3, pp. 18–26.

Commission Recommendation 2001/42/EC of 22 December 2000 concerning a co-ordinated Community monitoring programme for 2001 to ensure compliance with maximum levels of pesticide residues in and on cereals and certain products of plant origin, including fruit and vegetables, *Official Journal of the European Communities*, L 11, pp. 40–45.

6

Food contact materials

J. S. Baughan, Keller and Heckman LLP, Washington DC and J-P. Montfort, Keller and Heckman LLP, Brussels

6.1 Introduction

Since 1976, the Community has been working on the approximation of the laws of the member states regarding materials and articles intended to come into contact with food. In 1989, a new framework directive, calling for the adoption of specific Commission directives on ten different categories of food contact materials, was promulgated.

Today, only two main categories of materials (regenerated cellulose and ceramics) are subject to fully harmonized Community legislation, and the harmonization of the next category of materials (plastics) is still to be completed despite the fact that the first directive on plastics materials was adopted over ten years ago. In the meantime, one has to refer to the legislation in place in each of the individual European Union (EU) member states to determine the requirements applicable to those materials that are not subject to fully harmonized regulation at the EU level.

In the absence of fully harmonized EU legislation, the principle of 'mutual recognition' may be relied upon to ensure that materials and articles that comply with the regulation in place in one member state may freely circulate in the other parts of the EU. There is still some resistance to the application of the 'mutual recognition' principle in the food contact area, however, and this resistance, the slow progress towards harmonization in this area, and the continued use of different regulatory approaches by member states have maintained barriers to trade in food contact materials that are not justified for the protection of public health. Consequently, it is evident that a new approach towards harmonization in this area – one that prevents unnecessary barriers to trade while adequately safeguarding public health – is needed urgently.

One possible approach, recently implemented in the United States, would be to bring about a food contact notification (FCN) system in lieu of the 'positive list' system currently being pursued in the EU. An FCN system would be designed to allow the marketing of food contact materials after a defined review period, provided that the notifier has submitted to the authorities sufficient information to demonstrate that the intended use of the material is safe. The data required to be submitted to the authorities could be equivalent to the data requirements currently in place for petitions to the European Commission's Scientific Committee on Foods (SCF) to add materials to the directive governing plastic materials used in contact with food (the so-called 'Monomers Directive'); the improvement would be that the submitter would not be forced to wait for an indefinite amount of time for an amendment to a directive to permit the intended use of its material but, rather, would be able to go to market at a fixed time (e.g., 120 days) after the submission, unless the authorities were to object on safety grounds in the interim.

Of course, adoption of an FCN system in the EU is simply one suggestion for helping to resolve the current difficulty in navigating the sea of EU and national requirements relating to food contact substances, and is not the focus of this chapter. Our focus for the remainder of this chapter is as follows: we will first provide background information on the current legislative and other provisions that relate to the regulation of food contact substances in the EU. Next, we will discuss the principle of mutual recognition, its legal basis, and its applicability to food contact materials. We will follow this discussion with some practical examples of the way in which the existing food contact legislation and the principle of mutual recognition may be applied in specific situations. We will then complete our discussion with a look at the current trends in developing harmonized food contact legislation throughout the EU.

6.2 General EU legislation on food contact materials

The EU currently is in the process of 'harmonizing' legislation on food contact substances, principally by adopting directives that are designed to replace the existing national provisions of the member states. This work, however, is far from being complete.

The EU has begun by adopting a 'framework directive' governing all food contact materials (Directive 89/109/EEC), which provides the general safety criteria applicable to all food contact materials. The framework directive also provides for the adoption of specific directives covering ten individual categories of food contact materials. To date, only some of these specific directives have been adopted.

If a directive applicable to a particular product is in place at the EU level and has been implemented in the member states' national legislation, then the use of that product must comply with the directive. If an EU directive covering a particular product or application has not yet been promulgated, finalized or implemented into national law, then the use of the product must comply with the appropriate national laws of each of the EU member states, subject to the principle of 'mutual recognition', as described in section 6.6.

As indicated above, the EU Framework Directive 89/109/EEC[1] provides the general safety criteria applicable to all food contact materials.[2] More specifically, article 2 of this directive states that all food contact materials (1) must be manufactured in accordance with good manufacturing practices (GMPs), and (2) must not transfer their constituents to foodstuffs in quantities that could 'endanger human health' or bring about an unacceptable change in the composition of the food or its 'organoleptic' characteristics, i.e. they must not adulterate food. This directive has been implemented and is in place in all of the member states.

Article 3 of the framework directive provides for the adoption of more specific directives covering the following categories of food contact materials: (1) plastics, including varnishes and coatings, (2) regenerated cellulose, (3) elastomers and rubber, (4) paper and board, (5) ceramics, (6) glass, (7) metal and alloys, (8) wood, including cork, (9) textile products and (10) paraffin waxes and micro-crystalline waxes. These specific directives may contain 'positive lists' of the substances that may be used in contact with food, purity criteria, specific conditions of use, and specific and overall migration limits. They also may contain provisions permitting sampling and checking for compliance with established requirements, and other rules for the protection of public health. Based on the broad scope of the provisions these directives may contain, the Commission enjoys a fair degree of flexibility in determining the rules to be applied to specific food contact materials.

The framework directive also contains provisions relating to supplying information on food contact materials to consumers and professional users. In particular, it requires materials that are not already in contact with food to bear a 'fork and glass symbol', which is specified in Directive 80/590/EEC.

The European Commission currently is preparing an amendment to the framework directive, which would specify that the Commission is to adopt 'measures' instead of using the specific term 'directives' for specific food contact materials. This modification is intended to facilitate the application of new EU legislation in the member states. In particular, while directives must be implemented into the national laws of each individual member state, 'regulations' need not go through this additional procedural step. The Commission's current thinking is that the more specific directives promulgated pursuant to the framework directive often are so detailed that they are more like regulations than directives and, thus, should become effective in the member states under the doctrine of 'direct effect' even in cases in which the member states do not promulgate implementing legislation before the mandated deadlines. The use of the term 'measure' in place of 'directive' in this regard is thus considered by the Commission to be more accurate. The first amendment to the framework directive also would clarify or modify some of the labelling and other requirements of the directive. As discussed more fully below, however, the Commission is not yet ready to adopt general language on a 'Threshold of Regulatory Concern' within the directive, as supported by industry, which would be a key concept for the streamlining of legislation currently in place.

6.3 Legislation in place in the EU on specific food contact materials

While the framework directive requires the European Commission to adopt specific directives covering ten specific categories of food contact materials, to date only a few of these specific directives have been promulgated. Namely, directives on plastic materials, regenerated cellulose, ceramics and, to a limited extent, elastomers and rubber are currently the subject of specific directives addressing their food contact use. Further, the directive on plastic materials is not yet complete with respect to the additives used in the production of plastics; currently the directive is considered to be 'fully harmonized' only with respect to the list of monomers that are permitted in the production of food contact plastics.

In the meantime, one has to refer to the legislation in place in the member states to determine the requirements applicable to those materials that are not fully regulated at the EU level. Also, the principle of 'mutual recognition' may be used to ensure that materials and articles that comply with the regulation in place in one member state can freely circulate in the other parts of the EU, as discussed in section 6.6.

We discuss in detail below the requirements applicable to food contact plastics in the EU, since the EU directive on plastics (the Monomers Directive) is the most comprehensive of the directives on food contact materials adopted thus far. We note, however, that the Monomers Directive is limited in its scope and is not yet complete with respect to its 'positive list' of permitted plastics additives. Thus, plastic materials are still, to a great extent, regulated at the national level. This situation results in the need for expert analysis of the regulatory status of materials in many instances. The requirements applicable to other categories of food contact materials are also discussed below in a more general way.

6.4 Plastics materials and articles

6.4.1 The 'Monomers Directive' 90/128/EEC

The Monomers Directive 90/128/EEC[3] provides a complete positive list of permissible monomers for use in food contact plastic materials and an overall migration limitation (OML) that must be met for all plastics in contact with food, unless the use is subject to an exemption.[4] For most food contact materials, the overall migration must not exceed 10 milligrams per square decimeter of the article. Specific migration limits (SMLs) and quantitative limitations (QMs) (i.e. residual level limitations) have also been established for certain specific substances. Accordingly, all monomers intended for use in the production of food contact plastics must be listed on the Monomers Directive, and the final food contact article must meet the OML and any SMLs or QMs that have been established for substances used as components of the article.

The Monomers Directive also contains a list of substances that may be used as additives in the manufacture of plastic materials. These substances have been reviewed and considered safe by the European Communities Scientific Committee on Foods (SCF). The additives list, however, is not yet complete, and unlisted additives may still be used, provided that their use is demonstrated to be safe and meets any relevant requirements under applicable national laws.

It is important to note that the Monomers Directive does not apply to materials and articles composed of two or more layers, one or more of which does not consist exclusively of plastics, regardless of whether the layer in direct contact with food is exclusively plastic. (See Commission Directive 90/128/EEC, article 1, paragraph 4.) Thus, there currently is no legal requirement that each monomer or starting material used to produce a multi-layer article having one or more non-plastic layers be listed on the Monomers Directive. The regulatory requirements applicable to such products are limited to compliance with the general safety criteria of the framework directive and compliance with the applicable national legislation in place in the individual member states of interest. In this regard, however, a positive listing in the Monomers Directive for components of multi-layer materials having one or more non-plastic layers is relevant in that it indicates that such components have been evaluated by the SCF and determined to be 'safe' for use in contact with food, subject to any noted limitations.

In addition, the positive lists of monomers and additives that are included in the annexes to the Monomers Directive are not intended to include polymerization aids, nor are they intended to include substances used only in the production of surface coatings, silicones, epoxy resins, products obtained by means of bacterial fermentation, adhesives and adhesion promoters, or printing inks. Thus, these materials are also subject to the national laws of individual EU member states.

6.4.2 Other directives on food contact plastics

For reference only, we note that the only other EU directives related to food contact plastics are those concerning the testing of migration of the constituents of plastic materials and articles,[5] and those related to vinyl chloride monomer.[6]

6.4.3 National legislation covering food contact plastics

For substances that are not covered by a listing in the Monomers Directive, and in situations wherein the Monomers Directive does not represent fully harmonized legislation for the application of interest (if, for example, the unlisted material is an additive or an epoxy resin), then the national legislation in place in each of the individual member states must also be consulted to establish the status and confirm the safety of the substance for its intended use in food contact plastic applications. Eight of the fifteen EU member states (the United Kingdom, Germany, Austria, France, the Netherlands, Belgium, Italy and Spain) have some form of national 'positive list' of permissible substances for use in manufacturing food contact plastics beyond the required implementation of the EU directives. In Germany and the United Kingdom, however, these positive lists are not legally binding and other factors can be used to demonstrate that a given compound is safe. The remaining EU countries (Denmark, Finland, Greece, Ireland, Luxembourg, Portugal and Sweden) do not have any specific compositional requirements for food contact plastic materials other than those promulgated at the EU level, including the safety criteria established by the framework directive.

The laws of the member states that have 'positive lists' of permitted plastics additives are discussed briefly below.

Austria

Austria regulates food contact materials pursuant to its *Lebensmittelgesetz* (LMG) of 1975; some provisions of this law mirror those of the EU framework directive. Austria has also implemented the EU Monomers Directive and its amendments through a series of ordinances on plastics (*Kunststoffverordnung*), the first of which is the Ordinance No. 775 of 23 September 1994. Section 28 of the Austrian LMG of 1975 prohibits the marketing of food contact substances that are not approved or do not comply with the Austrian conditions of approval, or food contact materials and articles containing any such substances. 'Approved substances' include both substances that have been approved in Austria following petitions filed under the framework of the Austrian Food Act of 1951 and the LMG of 1975, and substances listed in the EU Monomers Directive (and, accordingly, the Austrian Plastics ordinances). Once a substance is 'authorized' in Austria, it may be used by any subsequent manufacturer, without any additional request or procedure, provided that such use complies with any specific conditions of use that may be prescribed.

Belgium

Belgium regulates food contact materials under the *Royal Arrêté* of 11 May 1992 on materials and objects intended for contact with foodstuffs (the '1992 Decree'), as amended. This decree governs the composition of food contact materials by means of 'positive lists' for various types of food contact materials, including the monomers, additives and aids to polymerization authorized for use in food contact plastics. The Monomers Directive and its amendments have been implemented into Belgian law. For some materials that are not covered by the Monomers Directive (e.g. additives and aids to polymerization), Belgian law also contains a 'positive' list of the substances that may be used to the exclusion of all others.

France

Food contact materials in France are regulated under a series of laws, decrees, *arrêtés* and circulars. Decree 73-128 of 12 February 1973 (the '1973 Decree') and a series of subsequent *arrêtés* and circulars, as reproduced in the *Recueil 1227* of the *French Official Journal*, provide, among other things, several positive lists of those starting substances and additives that are permitted for specified uses in food contact materials. These circulars, decrees and *arrêtés* are not organized according to type of product, so they all must be reviewed to determine whether a given substance is listed. As construed by French officials, the applicable French regulations constitute a 'positive list' so that the use of unlisted substances would require prior approval by French officials.

Germany

Germany regulates food contact materials pursuant to its law of 15 August 1974 on trade with foodstuffs, tobacco products, cosmetic agents and other articles ('*Lebensmittel und Bedarfsgegenständegesetz*' or 'LMBG'). Sections 30 and 31 of this law generally mirror the basic safety requirement set forth in the EU framework directive for food contact materials. In addition, Germany's regulation of 10 April 1992 on food contact materials implements the Monomers Directive. One way for a manufacturer to ensure that products that are not covered by the regulation of 10 April 1992 meet the LMBG's general safety requirements is to consider guidance contained in *Kunststoffe im Lebensmittelverkehr* of the *Bundesinstitut für gesundheitlichen Verbraucherschutz und Veterinärmedizin* (BgVV), also known as the 'BgVV Recommendations'. The BgVV Recommendations define specific positive lists of starting substances and additives, including reaction control agents, that are permitted for use in individual food packaging applications. Although they are not legally binding, the BgVV Recommendations are widely respected in Germany, and German manufacturers often insist that materials meet existing BgVV Recommendations. However, products whose safety can be demonstrated by other means are also equally compliant with German law.

Italy

Food contact materials in Italy are regulated under the decree of 21 March 1973 on hygienic requirements for packaging, containers, and utensils intended to be used in direct contact with food and substances for personal use ('the 1973 Decree'), as amended. This decree establishes rules for the authorization and control of objects intended to come into contact with food substances. Article 3, title I of the 1973 Decree stipulates that food contact materials must be prepared exclusively from components specifically listed in an attachment to the law for different categories of materials (such as plastic, rubber, regenerated cellulose, paper and cardboard, glass and stainless steel) and must otherwise comply with any conditions or limitations prescribed therein. The Ministerial Decree No. 220 of 26 April 1993 (the '1993 Decree') amended the 1973 Decree to implement the Monomers Directive into the laws of Italy.

The Netherlands

Food packaging materials are regulated in the Netherlands pursuant to the Decree of 1 October 1979 on packaging and articles of daily use (*Verpakkingen-en Gebruiksartikelen- besluit* (*Warenwet*)). This decree is implemented by the Ministerial Regulation of 25 January 1980 (the *Regeling verpakkingen en gebruiksartikelen* (*Warenwet*), as amended). These regulations are essentially a compilation of 'positive lists' for different types of substances, including plastics, that are permitted in the Netherlands for use in manufacturing food packaging materials. The *Warenwet* Regulations are structured in ten chapters that regulate plastics, paper and board, rubber, metals, glass, ceramics, textiles, regenerated cellulose, wood and cork, and coatings, respectively. As an example, chapter I on plastics applies to monomers, additives and aids to polymerization used in the production of food contact plastics.

Spain

Spain has implemented the EU directives on food contact materials into its own national law. Specifically, Spain has implemented the framework directive 89/109/EEC, as amended, by its Royal Decree 397/90 on materials and articles intended to be in contact with food and the Monomers Directive, as amended, by its Royal Decree 2207/94 on plastic materials and articles in contact with food (the '1994 Decree'). Spain's Resolution of 4 November 1982 contains a 'positive list' of the additives that may be used in food contact materials; however, this list has not been updated and it is understood that the Spanish authorities, in practice, take the position that Spain's implementation of the Monomers Directive supersedes this law with respect to plastic materials.

United Kingdom

Finally, the UK has adopted into its own laws the EU framework directive via the Materials and Articles in Contact with Food Regulations 1987 (SI 1987/1523), as amended, and the Monomers Directive via the Plastic Materials and Articles in Contact with Food Regulations 1998, as amended. Thus, monomers must be appropriately listed in the UK while, for additives, only the general safety criteria apply (i.e., additives that are not listed on the UK additives list may be used provided that they are safe). As a means of ensuring the safety of additives used in food contact plastic materials, the UK recognizes and relies upon the determinations and recommendations of the British Industrial Biological Research Association (BIBRA) and the British Plastics Federation (BPF) as listed in the *BIBRA/BPF Code of Practice*. The BIBRA Recommendations, however, are not legally binding and, thus, are of limited value. Consequently, other means, such as clearances in other countries, are often used to establish safety.

6.5 Other categories of food contact materials

The only other directives concerning materials and articles intended to contact food that have been adopted thus far at the EU level are those on regenerated cellulose film,[7] and on N-nitrosamines and N-nitrosatable substances from elastomer or rubber teats and soothers.[8] For other categories of materials, which are not yet the subject of specific directives, work towards the elaboration of common rules is being made in the forum of the Council of Europe. The Council of Europe (CoE) is a political organization that organizes cooperation between the governments of its member countries in a wide range of areas, including public health. (Currently, 41 European countries are members.) The CoE's Committee of Experts on Materials Coming into Contact with Food has the necessary expertise to develop guidelines, in the form of resolutions, applicable to the use of food contact materials. It has already developed resolutions on (1) colorants for plastics, (2) aids to polymerization for plastics, (3) ion-exchange and adsorbent resins, (4) surface coatings, and (5) silicones. The CoE currently is updating its coatings resolution, and is working on draft resolutions on food contact paper and paperboard, packaging inks and rubber, as well as guidelines on food contact metals and alloys, and wood and cork.

CoE resolutions are not legally binding; however, they are widely respected by governments and industry alike. Furthermore, there are groups who would like to have the CoE work on food contact materials translated into the national legislation in place in the EU member states. This approach would help to relieve some of the pressure on the Commission to move forward promptly with EU legislation on specific categories of food contact materials beyond plastics. This proposed path forward has created some controversy, however, as there is some concern that the procedures followed by the CoE are not as rigorous as those followed by the Commission with respect to completing scientific evaluations of materials and obtaining sufficient input from member states and from industry groups. Thus, progress on CoE resolutions may, for a time, be slowed.

6.6 The principle of mutual recognition

As mentioned above, if a directive applicable to a particular product is in place at the EU level and has been implemented in each of the member states' national legislation, then the use of the product must comply with the directive. Very few areas of food contact legislation have been fully harmonized, however, and progress towards this goal is extremely slow. In the meantime, in situations in which an EU directive covering a particular product or application has not yet been promulgated, finalized, or implemented into national law, then the use of the product must comply with the appropriate national laws of each of the EU member states, subject to the principle of 'mutual recognition'.

The principle of mutual recognition originates from the 'free movement of goods' principle, which is one of the fundamental principles of the European

Community. To achieve the free movement of goods, the authors of the Treaty Establishing the European Economic Community (i.e., the 'Treaty of Rome')[9] provided for the establishment of a 'Customs Union' and for the elimination of obstacles to free trade between member states. In particular, article 20 (formerly article 30)[10] is the basis for the mutual recognition principle, which requires the elimination of barriers on imports. It reads as follows: 'Quantitative restrictions on imports and all measures having equivalent effect shall ... be prohibited between member states.'[11]

Article 30 (formerly article 36) is the counterpart to article 28, stating that:

> The provisions of Articles 30 and 34 shall not preclude prohibitions or restrictions on imports, exports or goods in transit justified on grounds of public morality, public policy or public security; the protection of health and life of humans, animals and plants .... Such prohibitions or restrictions shall not, however, constitute a means of arbitrary discrimination or a disguised restriction on trade between Member States.

The application of articles 28 and 30 has evolved over the years, most notably in the landmark case of *Cassis de Dijon*,[12] in which the principle of mutual recognition was deemed to prohibit national legislation that does not allow the importation of 'products lawfully manufactured and marketed in another member state.' In essence, the principle allows for the legal importation and sale in a member state of products that are legally marketed in another member state even if the products do not comply with the specific regulatory requirements of the country of import. As interpreted by the European Court of Justice, this means that member states should allow products that are eligible for mutual recognition to circulate freely within their territory unless the member states are able to demonstrate, following an appropriate authorization procedure, that the product presents a danger to public health.

We note, however, that, when relying on mutual recognition as a basis for accepting a substance, national authorities generally are more likely to accept the marketing of substances with a positive listing in another member state, i.e. a substance that is explicitly included in a member state's positive list or one that has been the subject of a favourable evaluation by the SCF, the body responsible for evaluating food contact substances in the EU, as compared to a substance that has been marketed in another member state solely on the basis of a manufacturer's independent safety conclusion.

While the principle of mutual recognition as applied in other areas, such as the regulation of direct food additives, has been the subject of important case law by the European Court of Justice (ECJ), so far there has been no ruling by the Court in a case specifically involving food contact materials. Unfortunately, while there was a chance that this would change this year following an action filed by the European Commission against France in connection with France's Order of 9 November 1994 on rubber products in contact with foodstuffs, the case was dropped by the Court of Justice due to a procedural issue unrelated to the substantive issues raised in the case.

In August 1999, the European Commission decided to refer France's rubber decree to the Court of Justice[13] on the grounds that this legislation fails to allow the importation of rubber products authorized for use in another member state unless detailed rules on rubber products are applied by that State. Because detailed rules on rubber exist only in France, the Netherlands and Germany, the French rubber decree would, in essence, preclude companies marketing rubber products in other countries on the basis of determination of 'safety' instead of an explicit positive listing from marketing these products in France. The Commission, in its role of defending the free movement of goods principle enshrined in the EU Treaty, brought the case before the European Court of Justice in an effort to obtain a mandate for France to insert a 'mutual recognition' clause in its legislation. This case would have provided the Court with the opportunity to confirm, with respect to food packaging materials, its existing case law on direct food additives and other products. Unfortunately, however, the ECJ has dismissed the case on procedural grounds and, therefore, the substance of this case will not be argued in Court. Nevertheless, the conclusions of the Advocate General do, in substance, support the Commission's position on this case.[14]

Additional support for the use of the principle of mutual recognition for food contact materials may be found in the European Council Resolution on mutual recognition of 28 October 1999.[15] In this resolution, the Council encourages economic operators and citizens to make full use of the mutual recognition principle and invites the Commission to take measures to improve its application through information campaigns, guidebooks and brochures.

6.7 Determining compliance with EU food contact legislation: some practical examples

Having set the stage by discussing the applicable EU and national legislation governing food contact materials, we turn now to some practical examples of the way in which the relevant directives, national laws and regulations, and other useful concepts may be applied to specific products. Since, as discussed above, the most comprehensive EU-wide legislation applies to materials composed entirely of plastics, we will begin with this type of package.

6.7.1 Materials and articles made entirely of plastic

The first step in establishing a suitable status in the EU for a food package composed entirely of plastic is to ensure that each monomer used in the production of the package is listed in the Monomers Directive. As mentioned above, the Monomers Directive's positive list of monomers is considered to be exhaustive; thus, this part of the legislation is considered to be 'fully harmonized' in the EU. Unlisted monomers simply are not permitted for use in the production of food contact articles made entirely of plastic.[16] Once a positive listing for each monomer has been identified, the next step is to ensure that any limitations on the use of those monomers, such as specific migration limits, are met.

Turning then to additives, the simplest way to establish a suitable status for each additive used in a plastic formulation is to identify a listing in the Monomers Directive's positive list of permitted plastic additives and to establish that any listed limitations are met. Since the Monomers Directive and its amendments must be implemented into the national laws of each EU member state, a positive listing on the Monomers Directive ensures that the use of the additive is permitted in every EU country. If one or more additives is not listed, however, there may still be a basis upon which to establish a suitable status for the use of the substance. This is because, as discussed previously, the Monomers Directive's additives list is considered to be 'incomplete.' The additives list is still being developed; thus, this part of the legislation is not yet 'fully harmonized.' Consequently, with respect to additives, the national laws of the individual EU member states also apply.[17]

For member states that do not have a legally binding list of permitted plastic additives beyond the list that appears in the Monomers Directive, unlisted additives may continue to be used in food contact plastics, provided that such use is determined to be 'safe.' For countries that have national positive lists of permitted additives that go beyond the Monomers Directive listings, these positive lists must be considered. If each additive of interest is listed in each national positive list and meets any listed limitations, then the material may be used throughout the EU. The analysis does not stop there, however, in the (rather frequent) cases in which one or more additives is not listed. In such cases, it may be possible to rely upon the principle of mutual recognition as a legal basis upon which to market the material in member states wherein legally binding national additives lists do not include one or more of the additives of interest. To rely on the principle of mutual recognition, however, the additive must be safe for the intended use, and the product must first be lawfully marketed in another EU member state.

Finally, once a suitable status has been established for each component of a food contact plastic article, it is important to ensure that the article meets the overall migration limits set forth in the Monomers Directive.

Before turning to other types of food contact articles, it is worth noting that the regulatory situation in Germany with respect to plastic additives is unique. In particular, although Germany's BgVV Recommendations are not legally binding, they are widely respected and relied upon. Thus, although Germany may be considered to be a member state that does not have a 'legally binding' positive list of permitted additives beyond the list contained in the Monomers Directive, positive listings in the BgVV Recommendations are frequently desired to assure customers of the suitable regulatory status of an additive.

6.7.2 Complex plastic materials and articles

Plastics that are used in multi-layer articles that also contain non-plastic layers, such as multi-layer films that contain both plastic and foil layers, or laminated paper and plastic articles, are not subject to the positive list requirements of the Monomers Directive. These articles are subject only to the general safety requirements in place in all EU member states and the positive list requirements of those member states that specifically address these types of multi-layer materials. Thus, in many ways, the analysis one must undertake to establish a suitable status for these multi-layer products is similar to the analysis that applies to plastic additives.

For most member states, such articles may be marketed for use in food contact applications if it can be demonstrated that they are safe for the intended use and will not give rise to any taste or odour problems in the packaged food. For a few member states (namely Austria, France and the Netherlands), specific positive list requirements (and other requirements, depending upon the nature of the various layers) apply. Consequently, for these member states, the specific national requirements must be consulted. If a component of a multi-layer material is not listed on the relevant national list, then the principle of mutual recognition would need to be relied upon as legal support for marketing the material in the member state of interest. In this situation, the product would first need to be lawfully marketed in another EU member state.

Plastic articles that contain recycled content present unique regulatory issues. At the EU level, the directives governing food contact materials do not explicitly address the use of recycled materials in contact with food. Thus, food contact articles containing recycled materials are regulated at the national level.

We note that, while the Monomers Directive does not explicitly address the use of recycled materials in contact with food, the EU's 'Practical Guide for Users of European Directives' states as follows:

> Recently new procedures for obtaining monomers have been introduced e.g. by depolymerization of the finished articles already used. The Commission considers that these monomers can be used as starting substances for the manufacture of plastics intended to come into contact with foodstuffs, if they comply with the applicable EEC Directives. As regards the purity criteria of the mentioned monomers, see Directive 90/128/EEC [the Monomers Directive].

The Practical Guide is simply a guidance document and does not have the force of law. Nevertheless, the above-referenced provision makes clear that the Commission accepts the use of post-consumer recycled resins that are processed by depolymerization, provided that the finished monomers comply with the Monomers Directive and are of a suitable purity for the intended use.

As for EU member state requirements, most member states permit the use of recycled materials in contact with food, provided that the materials are demonstrated to be safe and suitable for the intended use. In this regard, although there currently is no EU directive that specifically addresses the use of recycled plastics in contact with food, the European Commission has sponsored a study on the criteria that should be considered for ensuring that recycled plastics are safe for such use. The study (referred to as the 'AIR Study'), which was conducted by the Agro-Industrial Research Programme and coordinated by the United Kingdom's Ministry of Agriculture, Fisheries and Food (MAFF), was accepted by the Commission in January 1998. The AIR study recommendations currently are relied upon by at least some member states as authoritative guidance on demonstrating the safety of recycled materials for use in contact with food.

Some countries, such as Belgium and France, require that data establishing the safety of particular recycled plastics must be submitted to and approved by government authorities. However, most member states do not require such submissions. Currently, the language of legislation in two member states, Italy and Spain, expressly prohibits the use of recycled materials in contact with food.

6.7.3 Other materials and articles

As for non-plastic articles, as detailed above, few materials are the subject of specific EU directives – most non-plastic articles are currently regulated solely at the member state level. Regenerated cellulose films are a notable exception. Council Directive 83/229/EEC, as amended, sets out detailed requirements relating to cellulose films, including a positive list of materials that are permitted for use in the production of coated and uncoated cellulose films. As this directive is considered to harmonize the legislation fully in this area, cellulose films marketed for use in food contact materials in any member state must comply with the requirements of this directive.

Rubber articles and ceramics are other types of packaging materials that are the subject of at least some legislation at the EU level. For ceramics, the EU legislation provides for lead limits and, for elastomers and rubber articles, the EU legislation provides for limits on the release of N-nitrosamines and N-nitrosable substances. As for compositional requirements for elastomers and rubber articles, in particular, national legislation still applies. Several member states, including Italy, France, Germany, Spain and the Netherlands, have specific positive lists of materials that may be used in the production of food contact rubber articles. (Again, Germany's list is a highly respected 'recommendation'.) In cases in which the use of an unlisted material is desired, the principle of mutual recognition again may be used as the legal basis for marketing the product in the 'positive list' country of interest. In this case, it is important to determine first that the product is safe for the intended use, then to lawfully market the product in another EU member state before marketing in the 'positive list' country. We note that, were such a situation to be brought to the attention of the authorities in France, the authorities may well not agree to the application of mutual recognition. This situation exists despite the action that the Commission has taken against France in connection with the French Order of 9 November 1994 on rubber products, discussed above.

As for materials such as metal, glass, and paper and paperboard, as well as products such as catalysts, coatings and adhesives used in the production of food contact articles, national laws still govern, subject to the general safety requirements of the framework directive. Although, as indicated above, the Council of Europe has prepared (or, in some cases, currently is preparing) resolutions to assist in establishing the safety and suitability of several of these types of products for use in food contact applications, the resolutions are not legally binding and, in some cases, progress on these resolutions is moving forward quite slowly. Thus, companies are in the unfortunate position of having to wade through the national legislation of every EU member state to determine which requirements apply to their products.

6.8 Future trends

While supporting the application of the mutual recognition principle to food contact materials that are not yet covered by harmonized directives, the Commission is continuing to develop and refine harmonized food contact legislation, still focusing mainly on plastics. In this regard, the Commission currently is preparing a sixth amendment to the Monomers Directive. This amendment, in addition to adding new monomers and additives to the positive lists, is expected to clarify the scope of the directive to exclude silicones (the Commission's view is that silicones are more properly defined as elastomeric materials and might be better regulated in a directive or resolution applicable to those types of materials). Further, it is expected to incorporate a provision to define better the relationship between SMLs and QM restrictions so that demonstrating compliance with an SML may be done more easily by calculation. In addition, the analytical tolerance listed in the Monomers Directive will likely be raised from 1 to 2 mg/dm^3 for aqueous foods. (Under annex I to the directive, a material that exceeds the overall migration limit by an amount not greater than the analytical tolerance is considered to comply with the directive.) These proposed changes to the Monomers Directive are consistent with some of the following trends in the Commission's thinking with respect to food contact regulation for the future.

First, the Commission has indicated that it would like to rely more heavily on the Council of Europe to proceed with harmonization efforts for products not covered by the Monomers Directive. As discussed above, the Council of Europe has already developed resolutions on colourants for plastics, aids to polymerization for plastics, ion-exchange and adsorbent resins, surface coatings, and silicones. Further, it is working on draft resolutions on food contact paper and paperboard, packaging inks and rubber, as well as guidelines on food contact metals and alloys, and wood and cork. The Commission hopes to be able to use these resolutions as the basis for directives on these categories of food contact materials. There have, however, been some concerns expressed in moving forward along these lines. In particular, some have cautioned that the

CoE procedures for developing resolutions are not as inclusive of member state and industry input, and do not employ scientific evaluations that are as rigorous as those employed by the Commission. Thus, progress on CoE resolutions and their possible use by the Commission as the basis for directives could be slowed considerably.

Second, the use of 'worst case' calculations in lieu of migration test data is gaining more widespread support in Europe both for demonstrating compliance with SML restrictions and for preparing dossiers requesting listings for new monomers and additives. The Commission recognizes the expense and difficulty that manufacturers and enforcement agencies face in attempting to analyse food or food-simulating solvents for the presence of plastic constituents and, as a result, is attempting to clarify existing legislation to promote the idea of analysing the plastic itself for the materials of interest and using the results to calculate potential levels of migration to contacted food.

Third, the idea of using 'food type' and 'material type' consumption factors to estimate potential exposures to food contact materials more accurately is being more carefully considered. Agreement on relevant consumption factors will be critical to the application of other useful concepts that are emerging in Europe, such as the concept of a 'threshold of regulatory concern' and the 'functional barrier' doctrine.

The EU system currently does not apply any material-type consumption factors or food type distribution factors to estimate potential dietary exposures to food contact materials. Instead, the SCF considers only the amount of a material that may migrate to food in determining how much toxicology data it must evaluate to support a determination of safety. In so doing, the SCF does not allow for the idea that only a fraction of all packaged food is packaged using a given material – a concept acknowledged by the US Food and Drug Administration (FDA) via its use of 'consumption factors' – nor does it allow for the idea that the typical human diet consists of a predictable combination of different food types (a concept acknowledged by the FDA via its use of 'food type distribution factors').

A significant amount of work has been done in Europe to promote the use of a 'fatty food consumption factor' in evaluating migration test results using fatty food-simulating solvents so that SML restrictions and required data packages for petitions may be more accurately assessed in situations involving contact with fatty foods. Work on developing data to support the use of other food-type consumption factors, and on what is being referred to in Europe as 'materials use consumption factors,' however, has not progressed to any great extent as yet. Nevertheless, the EU has indicated an interest in using consumption factors, and also has shown great interest in adding to its legislation concepts that relate to the use of such factors. In particular, the EU is working toward developing a 'threshold of toxicological concern', akin to the 'threshold of regulation concept' used by the FDA, to place a limit on the extent of toxicology data needed to support the safety of food contact substances that result in very low dietary exposures. Setting a 'threshold of toxicological concern' for food contact substances in the EU would help to simplify the regulatory process; however, to move forward with implementing this concept in a practical way, the Commission will need to resolve the consumption factor issue. By assuming that each material is used in contact with all types of food, and by neglecting to consider that the human diet consists of a distribution of different types of food, potential dietary exposures to individual food contact materials cannot be estimated with any reasonable degree of accuracy.

Work in developing a harmonized system for regulating food packaging in the EU previously has focused on formulating detailed positive lists of permitted materials. Now, the trend seems to be moving toward finding ways to adopt legislation and implement requirements in a more efficient and practical manner to lessen the time and financial burden associated with regulating these materials while continuing to ensure a high level of protection of public health. Who knows – perhaps the future will bring a trend toward the use in Europe of a pre-market notification system for regulating food contact materials similar to that recently adopted by the FDA.

6.9 Sources of further information and advice

JEAN-PHILIPPE MONTFORT, ' ''The Article 30 Solution'': an alternative to market food contact materials in the European Union, *Food and Drug Law Journal*, vol. 51, no. 1, 1996.

EUROPEAN COMMISSION, 'Practical Guide for users of EEC directives on materials and articles intended to come into contact with foodstuffs', 4 September 1998.

Synoptic Document, 'Draft of provisional list of monomers and additives used in the manufacture of plastics and coatings intended to come into contact with foodstuffs', 6 April 2001.

6.10 References and notes

1. Directive 89/109/EEC of 21 December 1988 on the approximation of the laws of the member states relating to materials and articles intended to come into contact with foodstuffs.
2. Article 1 of the framework directive makes clear that the directive applies not only to food contact materials and articles, but also to materials and articles that contact water intended for human consumption.
3. Directive 90/128/EEC of 23 February 1990 concerning plastic materials and articles intended to come into contact with foodstuffs (*OJ*), as amended by Directive 92/39/EEC of 14 May 1992, Directive 93/9/EEC of 15 March 1993, Directive 95/3/EC of 14 February 1995, Directive 96/11/EC of 5 March 1996, and Directive 1999/91/EC of 23 November 1999. A

'sixth amendment' to the monomers directive is expected to be published in the *Official Journal of the European Communities* later this year.

4. These exemptions apply to containers of certain volumes, sealing applications and articles that can be filled, and where it is impracticable to estimate the surface area in contact with food.
5. Directive 82/711/EEC of 18 October 1982, as amended, laying down the basic rules for testing migration of the constituents of plastic materials and articles intended to come into contact with foodstuffs, and Directive 85/572/EEC of 19 December 1985, as amended, laying down the list of simulants to be used for testing migration of plastic materials and articles intended to come into contact with foodstuffs.
6. Directive 78/142/EEC of 30 January 1978 on the approximation of the laws of the member states relating to materials and articles that contain vinyl chloride monomer and are intended to come into contact with foodstuffs and Directive 80/766/EEC of 8 July 1980 laying down the Community method of analysis for the official control of the vinyl chloride monomer level in materials and articles that are intended to come into contact with foodstuffs, and Directive 81/432/EEC of 29 April 1981 laying down the Community method of analysis for the official control of vinyl chloride monomer released by materials and articles into foodstuffs.
7. Directive 83/229/EEC of 25 April 1983 on the approximation of the laws of the member states relating to materials and articles made of regenerated cellulose film intended to come into contact with foodstuffs, as amended by Directive 93/10/EEC (*OJ*) and Directive 93/10/EEC of 15 March 1993 relating to materials and articles made of regenerated cellulose film intended to come into contact with foodstuffs, as amended (*OJ*).
8. Directive 93/11/EEC of 15 March 1993 concerning the release of the N-nitrosamines and N-nitrosatable substances from elastomer or rubber teats and soothers.
9. Treaty Establishing the European Economic Community, 25 March, 1957.
10. The provisions of the Treaty Establishing the European Community were renumbered as of 1 May 1999 pursuant to the Treaty of Amsterdam.
11. Article 29 (formerly article 34) of the Treaty provides a similar rule to prevent restrictions on exports.
12. Case 120/78, *Cassis de Dijon*, 1979 E.C.R. at 649.
13. Case C-230/99.
14. Conclusions presented by the Advocate M. Siegbert Alber on 7 November 2000.
15. Council Resolution of 28 October 1999 on mutual recognition (2000/C 141/02), *OJ* C 141/5 of 19 May 2000.
16. If a company wishes to produce a food contact material using a monomer that is not listed in the Monomers Directive, then a petition must be filed with the European Commission's Scientific Committee on Foods (SCF) to request a listing for the monomer of interest in the directive. This process, which involves a detailed review by the SCF of the data submitted and a

subsequent amendment to the Monomers Directive, can take years, considering in particular the time needed for the legislative process of adopting an amendment to the directive.

17. Similar to the situation for monomers, to include a new additive in the Monomers Directive's positive list, a petition must be filed with the Commission, and the directive must eventually be amended to include the new additive. Although the additives list is not considered to be 'fully harmonized,' at some point it will be considered complete. Therefore, it is advisable for companies to go forward with petitions to list new additives of interest in the Monomers Directive even though such a listing is not yet considered mandatory.

Part II

Informing the consumer

7

Labelling

D. A. Love, Consultant (formerly Ministry of Agriculture, Fisheries and Food, London)

7.1 Key principles

This chapter will deal with food labelling legislation of direct relevance to the consumer of food, i.e. the person who purchases food in a shop or a restaurant. It will not deal with the many rules relating to labelling of food at the pre-retail level. It will not, for example, look at the complex body of EU food hygiene rules even though some of the labelling provisions in them can result in information seen by the consumer in the shop.

The EU's food labelling legislation has developed piecemeal, without co-ordination and in the hands of different organisations with different objectives, but the majority of rules, including the main legislative text, are based on Articles of the Treaties which are concerned with the harmonisation of national legislation. When such harmonisation takes place, the EU's rules take precedence at the national level so that products from different member states are treated equally in every part of the Union. In other words the principle of non-discrimination between member states is established in the field of legislation 'occupied' by the EU rules. Where the rules are not comprehensive the EU text usually signals that national action is permitted but equally imposes on a member state the obligation to inform its partners and the EU Commission about its proposed action and often to obtain at least the Commission's approval. In this vetting procedure those informed will be motivated to promote the general Treaty principle of non-discrimination between member states.

For 20 years from 1979 to 1999 Council Directive 79/112/EEC was the principal food labelling legislation of the EU. In 2000 it and its amendments were consolidated into Directive 2000/13/EC. No substantive provisions were changed. The key principles are:

- to contribute to the smooth functioning of the internal market by removing impediments to free circulation of products and equal conditions of competition
- to inform and protect the consumer
- to prohibit the use of information that might mislead the purchaser.

These principles are embodied in rules for minimum mandatory information, or 'compulsory indications' in the language of the directive, which apply 'horizontally' to all foodstuffs put on the market and intended for sale. In other words, all foodstuffs must, unless exempted, carry labelling that contains certain minimum information and this list of mandatory information is set out in the directive. Any additional information, given either voluntarily by the supplier of the food or in response to specific rules laid down by the EU or by the member states, is subject to the principle that it should not mislead the purchaser. Any national rules are made subject to an EU vetting procedure laid down in the directive.[1]

Some EU food labelling legislation, depending on its subject matter, may have additional objectives. The Directive on Nutrition Labelling, 90/496/EEC,[2] is based first on the need to further 'the progressive establishment of the internal market', and second to assist the consumer in choosing foods appropriate to a healthy diet. The Regulation on Novel Foods, (EC) No. 258/97,[3] cites first the internal market, second the protection of public health and third the need to give the consumer 'necessary information'.

For a significant body of statutes, which lay down labelling information for foodstuffs subject to Common Agricultural Policy regimes, the key principles tend to be more concerned with the interests of food producers and of EU trade policy. They have, in fact, already been expressed in the Treaty of Rome, which listed the objectives of the Common Agricultural Policy as follows:

- increasing agricultural productivity
- thus ensuring a fair standard of living for the agricultural community
- stabilising markets
- assuring the availability of supplies
- ensuring that supplies reach consumers at reasonable prices.

Using the powers in the Treaty the EU has replaced national market organisations of agricultural products with common equivalents. Some of these products, e.g. fruits, vegetables and eggs, can be marketed without processing to consumers. Others, e.g. milk and meat, need some processing but are still covered by common market organisations. For many of these products the common organisations include rules on labelling for the retail stage and, therefore, for the consumer.

7.2 The organisation of EU legislation

The collection of rules that make up the EU's food labelling legislation can be organised as follows:

- the main food labelling directive, 2000/13/EC, and its associated subsidiary directives
- labelling rules in other horizontal food directives or regulations, currently covering nutrition labelling and novel foods
- labelling rules in marketing regulations adopted as components of Common Agricultural Policy management regimes for basic agricultural products, e.g. for eggs, apples, beef and even wine; and in directives dealing with the composition of certain processed foods
- miscellaneous rules such as regulations that establish schemes for food producers and suppliers to register their labels with the appropriate authorities so that they will be recognised throughout the EU, namely for protected designations of origin and traditional recipes or foods with 'specific character'
- disciplines on national labelling rule-making so that they do not conflict with the overriding EU principles of non-discrimination between member states and fair trade.

These rules were and still are generated by the different groups of officials in different parts of the Commission, responsible for agricultural, industrial, consumer protection or internal market policies. Some EU rules have been made in response to strong pressure from consumer lobby groups and increasingly the EU Parliament or certain member states. The motivations of the prime movers in these developments may have been on the one hand defensive or hostile to new technologies, irrespective of the evidence that these were properly assessed and controlled, and on the other frankly protectionist. The resulting EU rules often do not fit easily with existing rules or with the key principles. For example, it is now compulsory to inform the purchaser of the use of artificial sweeteners not only in the ingredients list (where all the other additives and ingredients are listed) but also in the legal name of the food. There is no safety reason for highlighting the use of approved sweeteners in this way and no requirement to treat other additives similarly. It has also recently been made compulsory to make a special declaration on the label when packaging gases are used in containers. The gases are used to extend the shelf-life of products but, because they are not ingredients of the food, they will not be listed in the ingredients list. There are no safety concerns about the use of these gases and the meaning of the declaration for the vast number of purchasers is obscure.

7.3 EU legislation and Codex standards

With some exceptions this collection of EU food labelling rules does not differ markedly from the standards and guidelines of the Codex Alimentarius. The Codex Alimentarius is the body of food standards which has commanded the consensus of countries which have participated and negotiated in regular meetings of the Codex Committees established by the UN's Food and Agriculture Organisation and World Health Organisation. Food labelling is the subject of well-organised, annual meetings in Canada of the Codex Committee on Food Labelling, which is a powerful committee with responsibility for setting horizontal standards and guidelines and for vetting the more specific labelling rules proposed by other Codex committees responsible for product sectors. There has hitherto been broad international consensus on what labelling rules should cover. Briefly, they should require that the food is correctly identified and described, including all its ingredients, if necessary. Legislators have taken the view that it is the nature of the food as purchased and consumed that is of prime concern to the purchaser. Other types of information, e.g. production methods, are treated as voluntarily made claims which are controlled only by the general obligation not to mislead. However, if certain claims become widespread, rules are usually devised to define such terms in order to protect consumers from fraud and *bona fide* producers from unfair competition.

However, there are signs that these traditional labelling principles may no longer be enough and that the international consensus is changing. It is being challenged increasingly by self-styled 'public interest' groups which are seeking mandatory information on labels about such matters as geographical origin, production methods, treatment of animals and use of new technologies. The labelling of foods produced from genetically modified crops or from processes involving at some stage a genetic modification is currently the main focus of this debate. The reports of recent meetings of the Codex Committee on Food Labelling, which has discussed the labelling of foods produced by modern biotechnology, reveal clearly the serious lack of consensus on this issue.[4] Most member countries, with the strong support of some pressure groups, now accept that information must appear on the label about the use of the technology, irrespective of its effect on the food as purchased. A few see no need for such an indication, unless the technology has changed the food in some way.

7.4 The main requirements for prepacked foods

The EU's main food labelling directive, 2000/13/EC applies to all foodstuffs delivered to the consumer, whether at the retail stage or in catering establishments. It imposes two general rules: that labelling, presentation and advertising should not mislead the purchaser to a material degree, with some helpful elaboration on how that might occur; nor should they carry any medicinal claim about a foodstuff, i.e. a statement that it has the property of preventing, treating or curing a human disease or a statement with any reference to such properties. The directive lays down the following categories of information which must appear on labelling:

- the name of the food
- the list of ingredients
- the quantity of certain ingredients

- the net quantity
- the date of minimum durability
- any special storage conditions or conditions of use
- the name and address of the manufacturer, packager or seller
- place of origin, if omission of such information would mislead
- any necessary instructions for use
- alcoholic strength by volume for beverages containing more than 1.2% by volume.

Of these categories some must appear in the same field of vision on the label: the name, the net quantity and the date-mark, plus the alcoholic strength for alcoholic drinks. Most of the rest of the directive lays down more detailed rules for several of these categories. It also prevents member states from imposing other mandatory information unless this has been authorised at EU level.

7.4.1 The name of the food

If a name of a food is laid down in EU legislation, it must be used for the foodstuff in question. Failing that, the name prescribed in the member state where the product is marketed must be followed. If there is none, the supplier must use either a customary name or a description of the foodstuff that is clear enough to convey to the purchaser its true nature. The legal name in the member state where the food is produced is given equal status to the legal name of that food in the member state where it is marketed, but if together with the other label information it would not enable consumers to know the true nature of the food, then descriptive information must accompany the name or in exceptional cases the name cannot be used. The legal name cannot be the brand or fancy name nor the trade mark used on the label. If the food has been treated or its physical condition has been changed, e.g. dried, concentrated or frozen, this must be indicated in the legal name, if omission of such information would confuse. For one treatment, irradiation, the precise terminology to be used has been laid down.

7.4.2 The list of ingredients

The directive sets out fairly detailed rules on ingredients, with the help of three annexes. The main provisions in the article text define ingredients and require that they be listed in descending order of weight as recorded at the time of their use in the manufacture of the food, and that they be given specific names. A specified list of foodstuffs, most notably alcoholic drinks containing more than 1.2% alcohol by volume, need not carry ingredients lists. 'Ingredient' is any substance that is used in the manufacture or preparation of the food and is still present in the finished product. It therefore includes additives but not those that are used as processing aids, solvents or media for other additives or flavourings nor those that may be present in the final product but serve no technological function in it. Specific rules deal with added water, concentrated or dehydrated ingredients and mixtures of fruit, vegetables, herbs and spices. Annexes I and III provide that flavourings and categories of certain ingredients may be described in the ingredients list with general rather than specific names. Annex II lays down the additive category names that must be used. A compound ingredient may be included in the list under its own name and in terms of its overall weight, provided that its name is immediately followed by a list of its own ingredients, unless it constitutes less than 25% of the product or is a food exempt from ingredient listing. The 25% rule exemption does not apply to those additives in a compound ingredient which serve a technological function in the final food, which must all be listed.

7.4.3 The quantitative declaration of ingredients (QUID)

Article 7 of the directive contains the recently agreed, so-called QUID rules relating to quantitative declaration of ingredients. All ingredients that appear in the legal name, or are usually associated with the food, or which are emphasised on the label, or which are essential to characterise a food and to distinguish it from products with which it might be confused, should be quantified. Exemptions apply to foods already covered by quantitative information requirements in other EU legislation, and remove the obligation to quantify for ingredients used in small quantities for the purposes of flavouring, for the mixtures of fruits, vegetables, herbs and spices that do not need to be separately identified in the ingredients list, and for ingredients that while appearing in the name of the food, are judged not to influence the choice of the consumer as far as their quantities are concerned. The quantity must be shown as a percentage; it must relate to the 'mixing bowl' or preparation of the food, not the finished product; and it must appear next to the name of the food or to the name of the ingredient in the ingredients list. (See also section 7.4.7.)

7.4.4 Net quantity

The rule on net quantity applies only to pre-packed foods and establishes the principle that net quantity must be expressed in metric units of volume (for liquids) or mass (for other products), unless EU or national provisions lay down that some other type of quantity is required. More detailed requirements are laid down in Council Directives 75/106/EEC and 76/211/EEC and Commission Directive 78/891: that 'nominal weight' or 'nominal volume' must be shown on the package, that the contents must on average not be less than the nominal quantity and that a small 'e' must be placed in the same field of view as a guarantee that the information meets the requirements of the directives.[5] For solid foodstuffs sold in liquid media (as defined) the drained net weight must also be shown.

7.4.5 Minimum durability

There are two detailed articles in the directive on the important issue of date of minimum durability, which is defined as the date until which the foodstuff retains its specific properties when properly stored. It must be indicated by use of the words 'best before' and the date itself, which can be shown in several ways depending on the food's shelf-life. Certain foods and drinks are exempted from date-marking. Foodstuffs which are microbiologically highly perishable must carry a 'use by' in place of the 'best before' indication, and the form of the date is specified.

7.4.6 Other requirements

The remaining articles of the directive deal with the 'competence' of the member states in relation to some important general issues. One permits them not to apply some or all of the minimum mandatory information to foods sold without prepackaging or those sold in fancy packaging. In the UK there are very significant differences between the information required for prepacked foods and for those sold loose, e.g. at greengrocery or delicatessen counters, or those sold 'prepacked for direct sale', i.e. wrapped at the retail stage usually following the customer's order. Another article obliges member states to require that the minimum mandatory information on labels should be in a language easily understood by the purchaser, although they may also stipulate that certain EU languages must be used.

7.4.7 Recent supplementary provisions

This directive is supplemented by three further directives, adopted in 1994, 1996 and 1999.[6] The two earlier ones use the power in article 4(2) of the principal directive to add to the minimum mandatory information for certain foodstuffs. The first requires that the words 'packaged in a protective atmosphere' should appear on the labels of foodstuffs whose shelf-life has been extended by means of packaging gases. The second requires mention of sweeteners both in the ingredients list and in the name of the food, as well as other wording when certain sweeteners are used. The most recent subsidiary directive modifies the QUID rules in the light of further consideration about how they could be implemented (an exercise largely driven by the UK, which accepted the original rules with great reluctance because of their vagueness and subjectivity). Now volatile ingredients must be indicated on the basis of their proportion by weight in the final product, and concentrated or dehydrated ingredients may be indicated in their whole form if that is how they are consumed. Unnecessary quantifications of sweeteners, vitamins and minerals are now set aside. The EU has also adopted guidelines, again largely drafted by the UK, on the practical application of the QUID rules.

This body of generally applicable food labelling requirements is very similar to the General Standard on the Labelling of Prepacked Food of the Codex Alimentarius.[7] The most important difference is that the Codex Standard does not include mandatory requirements for the quantitative declaration of ingredients. Nevertheless, there is no reason that Codex should not adopt these requirements. They conform to the internationally accepted principles of informing the consumer about the nature of the food as purchased and of enhancing equal conditions of competition between food suppliers. Quantitative declaration of ingredients is an important additional piece of information about the composition of many so-called compound foods, which allows consumers to compare competing products which may appear similar. It becomes less important if the products in question are all made to the same recipe because legislation defines the composition. QUID, when adopted, facilitates the repeal of sometimes rather old-fashioned 'recipe law', although the attachment of some legislators and indeed some industry sectors to 'their' laws should not be underestimated.

7.5 Nutrition labelling and claims

Council Directive 90/496/EEC on nutrition labelling[2] established a standardised format within the EU for quantitative nutrition information on foods. Its legal basis is Article 100A of the Treaty of Rome, making it an internal market measure and this is reinforced by references in its preamble to enabling consumer choice and avoiding technical barriers to trade. However, there is also recognition that there are other dimensions to this particular subject; the growing public interest in the relationship between diet and health and the improvement of nutrition in order to assist the European programme against cancer. Therefore, the legislation is presented as providing for a voluntary, standardised, simple, easy to understand scheme which should be introduced gradually and then reviewed and amended in the light of its operation.

The rules are potentially applicable to all food and drink except natural mineral waters or other waters intended for human consumption. The directive provides that nutrition labelling as defined is optional but it becomes mandatory if a nutrition claim as defined appears on labelling, presentation or advertising of a product. Nutrition labelling means any information on the label that refers to the energy value of the food, or to protein, carbohydrate, fat, fibre, sodium or to other minerals or vitamins listed in the annex to the directive. Nutrition claim is defined as any representation and any advertising message that states, suggests or implies that a foodstuff has particular nutrition properties relating to energy value or to its nutrients. The directive goes on to prohibit any nutrition claims that do not relate to energy, the nutrients mentioned above or the substances that belong to or are components of a category of those nutrients, e.g. starch as a type of carbohydrate.

Whenever nutrition infomation is given it must be given in one of two basic formats, preferably in tabular form with numbers aligned:

1. Group 1, which consists of energy value in kJ or kcal and the amount of protein, carbohydrate and fat in grams.
2. Group 2, which consists of those plus sugars, saturates, fibre and sodium, also in grams.

If a nutrition claim is made for any of the additional Group 2 nutrients, the Group 1 format cannot be used. Both formats could be expanded to include additional components, namely starch, polyols, mono-unsaturates, polyunsaturates, cholesterol and the vitamins and minerals listed in the annex; but if information about mono-unsaturates, polyunsaturates and/or cholesterol is given, the amount of saturates must also be specified. The effect of these rules is to prevent the voluntary addition of information about certain nutrient components that are not mentioned in the directive. The most important of these in terms of current interest is trans-fatty acids. If a claim about trans-fatty acids is made, the quantity must be shown in the nutrition labelling format used; if no claim is made, strictly speaking no information about trans-fatty acids can be given.

The directive goes on to lay down definitions of certain nutrients, energy values for the main nutrients and further rules on the formats. For example, the quantities given must be expressed per 100 gm or per 100 ml of the product and may be expressed also per serving or per portion of the product. They must be given in relation to the food as sold but may alternatively be in relation to the food as consumed, provided sufficiently detailed preparation instructions are given. The quantities of vitamins and minerals must be given not only per 100 gm/ml but also as percentages of the recommended daily allowances (RDAs) specified in the annex. The annex itself contains an important little statement, the effect of which is to prevent *in most cases* any quantitive information on vitamins and minerals unless the food contains 15% of the relevant RDA.

Although the directive is a lucidly drafted set of instructions to member states, the UK's experience of implementing it was dogged with uncertainty about its practical operation. Food suppliers and food law enforcers have needed a great deal of advice and guidance, and the Ministry of Agriculture, Fisheries and Food (MAFF) eventually produced detailed guidance notes in a question-and-answer format in order to elucidate the rules and tackle issues of detail such as calculations of vitamins, energy conversion factors for novel nutrients, and the understanding of phrases such as 'generally speaking'.

Since the adoption of the nutrition labelling directive the Commission has expressed its intention of making proposals for the harmonisation of member states' rules on nutrition claims. There have been several attempts but they have never got beyond internal consultation drafts, because of differences of opinion within the Commission, some resistance from industry and very different attitudes to such claims in member states. In the UK the control of such claims is patchy. Apart from the general rules that a claim should not mislead the purchaser and that medicinal claims are prohibited, there are legislative criteria for claims about energy, protein, vitamins, minerals and cholesterol. The UK's Food Advisory Committee had also recommended criteria for claims about fat, saturates, sugar(s), salt/sodium and fibre, which have been adopted by many manufacturing and retailing companies but not by all, because of serious disagreement about the applicability of the recommended criterion for the 'low fat' claim to certain types of products. The UK has acknowledged for some time that this situation needs to be improved but has held back, first in the expectation of proposals from the Commission and later because of promising progress on developing guidelines in the Codex Committee on Food Labelling. That Committee now has agreed guidelines for criteria for a longer list of claims than are covered by the UK's legislation and recommendations combined. The UK's criteria have been amended as a result of consultations and adopted formally. They and the guidelines on nutrition labelling are now available from the Food Standards Agency.[8]

7.6 Specific labelling requirements in food composition directives

This group of directives was put together mainly during the 1970s, when the EEC embarked on the attempt to harmonise national laws on specific foodstuffs. All the directives were based on Article 100 of the Treaty of Rome and most of them were also based on Article 43 because they were controlling foods subject to the common organisations of agricultural markets. Nine food sectors were covered:

- cocoa and chocolate products
- coffee extracts and chicory extracts
- certain preserved milks
- fruit juices and similar products
- honey
- certain sugars
- jam
- caseins
- erucic acid.

The first seven of these have been reviewed by the EU in order to simplify and update them in the light of the horizontal rules adopted on labelling and additives. At the time of writing the directives on cocoa and chocolate products and on coffee and chicory extracts had been finalised.[9] The others were still under negotiation in the EU's institutions. The following paragraphs give examples of the kind of provisions that the current seven directives contain.[10]

The preambular clauses always base the need for harmonised rules on the fact that national laws exist that define the products in question and impose labelling conditions, on the likelihood that these national laws will constitute a barrier to the free movement of goods and on the assertion that they therefore have a direct bearing on the establishment and functioning of the common market. Council Directive 73/437/EEC on the approximation of the laws of the member states concerning certain sugars intended for human consumption also justifies its provisions by the need to ensure the smooth running of the common organisation of the market in sugar. This directive, in common with some of the others, makes no reference to the consumer, while others in the seven do contain justifications about protecting and informing consumers and about laying down conditions that take account of consumer requirements.

All the directives in this group define the products that they cover and assign each product a name which can only be applied to a food meeting the product's definition and moreover which must be used for that food in trade. Those directives adopted before the principal food labelling directive, 79/112/EEC, also list the information that must appear on the labels of the foods in question, for example, the name of the manufacturer, packer or seller in the EU. Many require net weight to be shown and in some sale quantities are prescribed. Depending on the product in question other descriptions or types of information are made compulsory.

Member states are generally not permitted to make any other information mandatory but there are many exemptions, indicative of the difficulties that arise when vertical harmonisation is attemped. Product names specific to one or a few member states have to be listed, e.g. evaporated milk is allowed to be a reserved name in the UK and Republic of Ireland. Some of the directives permit member states to retain their national rules requiring the origin of products to be stated, provided that they are not applied to products of EU origin. National rules on the addition of vitamins to preserved milks and fruit juices are permitted. National rules on how the information should be presented on the label are prohibited but member states are allowed to specify the languages that must be used.

7.7 Specific labelling requirements in CAP marketing regulations

CAP marketing regulations which contain labelling provisions of direct interest to consumers deal with eggs, poultry, fruit and vegetables, olive oil, beef, milk and milk products and spreadable fats; and also with wines and some spirit drinks (because of the use of agricultural produce, namely grapes and cereals, in their production). All these rules are devised by officials in the Agriculture Directorate General and negotiated within the Council of Agricultural Ministers or in the committees charged with the day-to-day management of the common markets in agricultural produce. The preambles of the regulations generally state that the aims are to promote the stability of the agricultural market in question, safeguard the interests of producers and help consumers identify and distinguish foods that may be confused or foods that may differ in quality.

The labelling rules have been devised to meet the different marketing conditions of each foodstuff and it is not possible to treat the foods as a homogeneous group, unlike the foods covered by the food composition directives described in the previous section. The rules range from simple provisions in the olive oil regulations, which are concerned mainly with defining the meaning of descriptions such as olive oil and virgin olive oil, to detailed rules on eggs or on the use of dairy product names.[11]

One important similarity between some of the regulations is the grading of the foods they control by quality. This applies to fruits and vegetables, eggs and poultry and such classifications are justified as contributing to the improvement in the quality of the foods in question, facilitating trade and making it easier for consumers to distinguish between products of different quality. The regulations require that labels should indicate the quality class. Another similarity between some regulations is the requirement to show country of origin. This applies to fruits and vegetables and is justified as an important means of conveying to consumers the characteristics of these foods. Poultry meat from countries outside the EU must show its origin. The recently agreed regulation on the labelling of beef indicates that country of origin will become compulsory in due course. Production methods for eggs and poultry are of interest to many consumers and the regulations duly define the chief features of such systems as free range or barn reared and require that products marketed under those terms should conform to the definitions. Other information prescribed or controlled by the marketing regulations include date-marking, weight, the condition of the food (i.e. fresh, frozen), variety, even price indications and the size of lettering, some of which is clearly not necessary, given the existence of the general rules in 79/112 (now 2000/13).

Finally, most of the regulations set out names and descriptions that should be used for the foods when they are marketed to the consumer. This feature is most developed in the rules covering milk products and spreadable fats. Council Regulation (EEC) No. 1898/87 on the protection of designations used in the marketing of milk and milk products has as its stated objectives that the natural composition of these foods should be protected in the interests of EU producers and consumers by means of appropriate labelling, and that confusion between milk products and competing products, e.g. margarines, should be avoided. The related Council Regulation (EEC) No. 2991/94 laying down standards for spreadable fats also aims to help consumers distinguish between the products covered, which may be comparable in terms of fat content but differ in terms of the types of fats used, i.e. milk, animal or vegetable. These objectives are achieved by detailed rules on names and descriptions of the products in question. Regulation 1898/87 defines milk and milk products (whey, cream, butter, buttermilk, butteroil, cheese, yoghurt and some others), reserves the names and prohibits their application or the use of related terms, such as 'dairy', to other products. Regulation 2991/94 distinguishes spreadable fats by name and by fat content and the names are reserved. Both regulations recognise that some of the reserved names have traditionally been used for completely different foods, e.g. cream cracker, coconut milk, and they provide for exemptions for the use of reserved names in 'products the exact nature of which is clear from traditional usage and/or when the designations are clearly used to describe a characteristic quality of the product', e.g. cream of tomato soup. However, use of the term 'butter' in what are called composite products is permitted only if the products are recognised in the regulations, which have been amended to list permitted names, such as brandy butter, and to specify their minimum milk fat content.

7.8 Novel foods and genetically modified foods: labelling rules

The regulation on novel foods, including genetically modified foods, (EC) No. 258/97,[3] subjects novel foods, as defined, to a pre-marketing approval procedure, resulting either in no objection to a member state's approval or an 'authorisation decision' at EU level for each approved novel food. The authorisation is likely to require the supplier of the novel food to include labelling information on their product in addition to the information required by existing EU labelling rules. This additional information is set out in article 8 of the regulation as follows:

(a) the presence of genetically modified organisms;
(b) the presence of material in a novel food which is not present in an existing equivalent food and which gives rise to ethical concerns;
(c) the presence of material in a novel food which is not present in existing equivalent food which may affect the health of certain sections of the population;
(d) a characteristic or property of the food, such as composition, nutritional value or effect, or intended use, which renders it no longer equivalent to an existing food. In this case equivalence is to be assessed scientifically and the assessment will take account of accepted limits of natural variations of the characteristics assessed. Also in this case the labelling will identify the modified characteristic or property and the method by which it was obtained.

Because of its definition of novel foods, the provisions of this regulation could not be applied to novel foods and genetically modified foods that were already on the EU market. Two such examples were foods that resulted from the application of genetic modification techniques to soya and maize. Commission decisions of April 1996 and January 1997, made under the EU's directive on the deliberate release into the environment of genetically modified organisms, had approved genetically modified soya beans with increased tolerance to the herbicide, glyphosate, and genetically modified maize with both insecticidal properties and increased tolerance to the herbicide, glufosinate ammonium. These raw products were on the EU market and, although they themselves contained live genetically modified organisms, food products derived from them, e.g. oil and flour, do not. Nevertheless there was considerable concern about them among some member states, who did not regard the existing food labelling rules as sufficient for them. In May 1998 Council Regulation 1139/98 specified the wording required on labels when foodstuffs contained ingredients derived from genetically modified soya and maize, if they contained protein or DNA material resulting from genetic modification. If they did not contain such protein or DNA, even though they might have been obtained from genetically modified soya beans or maize, these labelling rules did not apply. The regulation also provides for the construction of a list of processed soya and maize products not subject to the additional labelling requirements. When it appears (probably including refined oils and starch derivatives) it will help companies by giving greater certainty about the products affected by the regulation. Another area of rule-making is signalled in the regulation's preamble, namely setting a *de minimis* threshold for the presence of DNA or protein resulting from genetic modification in order to take account of adventitious contamination.[12] This was subsequently set at 1 per cent and the rules were extended to additives and flavourings produced from genetically modified organisms from all sources, not just soya and maize.[13]

Regulation 1139/98 is important because it gives legislative form for the first time to the EU's developing ideas on what 'equivalence' means for labelling purposes in the context of genetic modification. Internationally, the concept of 'substantial equivalence' is used in relation to the safety assessment of genetically modified foods. 'Substantial equivalence' has been described in the WHO/FAO paper, 'Biotechnology and Food Safety'.[14] Its basic concepts are repeated almost verbatim in Article 8 of the novel foods regulation (258/97). However, the EU's refinement of these concepts in regulation 1139/98 is regarded by the other main international regulators (in the USA and Canada) as requiring labelling in more cases than may be justified.

7.9 Future developments

The Commission's White Paper on Food Safety, published in January 2000, reveals that no major changes are planned in the existing principles or organisation of labelling rules. A commitment is made to propose the repeal of the 25% rule exemption (see section 7.4.2 above). It will also consider further specific rules on ingredient listing so that known allergens can be identified. Moreover, it will consider proposing specific rules to control claims about the presence, absence, level or effect of nutrients and updating the Nutrition Labelling Directive. The Commission also commits itself to clarifying the rules governing the labelling of novel foods, particularly on the traceability and labelling of products derived from genetically modified organisms.

These are obviously the areas in which we should expect to see more activity in future and they are similar to developments in the Codex Committee on Food Labelling. Recent changes to that forum's standard and guidelines include a modification to the 25% rule to zero for a specified list of foods which have been demonstrated to cause hypersensitivity, and established criteria for the most commonly used nutrition claims. It has also embarked on guidelines for health claims. There is also an outstanding proposal from the Commission to extend ingredient listing to all alcoholic drinks.

Clearly, more radical changes or new directions will not be seen in EU food labelling legislation. The Commission does not seem to be keen to lead the way and in the absence of a commitment from it, the realistic prospect for labelling legislation is to continue to develop in a piecemeal and uncoordinated fashion.

7.10 References and sources of further information

NOTE: the abbreviation *OJ* is used below for the *Official Journal of the European Communities*, which can be consulted in specialist libraries and directories.

1. Directive 2000/13/EC, of the European Parliament and of the Council, *OJ* No. L109 of 6 May 2000, p. 29.
2. Council Directive 90/496/EEC, *OJ* No. L276 of 6 October 1990, p. 40.
3. Regulation (EC) No. 258/97 of the European Parliament and the Council, *OJ* No. L43 of 14 February 1997, p. 1.
4. See, for example, reports of the twenty-fourth and twenty-fifth sessions of the Codex Committee on Food Labelling, FAO, Rome, 1996 and 1997.
5 Council Directive 75/106/EEC, *OJ* No. L42 of 15 March 1975, p. 1; Council Directive 76/211/EEC, *OJ* No. L46 of 21 February 1976, p. 1; Commission Directive 78/891/EEC, *OJ* No. L311 of 4 November 1978, p. 21.
6. They are Commission Directive 94/54/EC, *OJ* No. L300 of 23 November 1994, p. 14; Council Directive 96/21/EC, *OJ* No. L88 of 5 April 1996, p. 5; and Commission Directive 99/10/EC, *OJ* No. L69 of 16 March 1999, p. 22.
7. General Standard of the Labelling of Prepacked Foods of the Codex Alimentarius, FAO, Rome, 1998.
8. Its address is Aviation House, 125 Kingsway, London WC2 6NH.
9. Directive 2000/36/EC of the European Parliament and the Council (cocoa and chocolate products) *OJ* No. L197 of 3 August 2000, p. 19; and Directive 1999/4/EC etc. (coffee extracts and chicory extracts) *OJ* No. L66 of 13 March 1999, p. 26.
10. These Directives are: Council Directive 76/118/EEC (preserved milks), *OJ* No. L24 of 30 January 1974, p. 49; Council Directive 75/726/EEC (fruit juices, etc.), *OJ* No. L311 of 1 December 1975, p. 40; Council Directive 74/409/EEC (honey), *OJ* No. L221 of 12 August 1974, p. 10; Council Directive 79/693/EEC (jams), *OJ* No. L205 of 13 August 1979, p. 5; and Council Directive 73/437/EEC (sugar products), *OJ* No. L356 of 27 December 1973, p. 71.
11. The list of regulations covering this group of foodstuffs is very long and amendments appear frequently. Specialist libraries will be able to provide access to the texts and the relevant databases.
12. Council Regulation (EC) 1139/98, *OJ* No. L159 of 3 June 1998, p. 4.
13. The ammendments are in Commission Regulation (EC) No. 49/2000, *OJ* No. L6 of 11 January 2000, p. 13 and Commission (EC) No. 50/2000, *OJ* No. L6 of 11 January 2000, p. 15.
14. 'Biotechnology and Food Safety', *FAO food and nutrition* p. 61, FAO, Rome, 1996.

8

Nutrition information

L. Insall, Food and Drink Federation, London

8.1 Introduction: key issues in presenting nutrition information

Increasing interest in the relationship between diet and health has led to an ever sharper focus on the nutritional aspects of the food supply. Interest in nutrition, in respect of both total diet and individual foods, is second only to concern about food safety, and is sometimes confused with the safety aspects of the food supply. Some commentators blame the nutritional content of the diet for a wide range of health problems such as obesity, cardiovascular disease and certain cancers, that is, long-term chronic diseases as opposed to the type of short-term acute illnesses usually associated with food safety problems.

Nutrition information is, therefore, an aspect of a very broad debate, often highly politicised, about the nutritional quality of the modern food supply, and specifically about the contribution to the modern diet, and therefore the health of the population, made by the food manufacturing industry. In terms of issues, the provision of nutrition information ranks very high in the diet and health debate. The United Kingdom has possibly been more absorbed by this subject than most other European member states, a reflection, perhaps, of the cultural attitude towards food as fuel and the growing obsession with fitness and body form in a population inclining towards obesity. Where food has traditionally been enjoyed as one of life's great pleasures, notably France, the most important factor is what the product tastes like, not the 'vital statistics' of its content.

Nevertheless, the link between food and health is now acknowledged as a matter of considerable importance to the EU as a whole and is the subject of a major review, with the ultimate aim of introducing dietary guidelines throughout the EU. The UK experience is therefore likely to be mirrored to a greater or lesser extent in other member states and UK practices and consumer research are used here as an illustration of issues of wider relevance on which other member states may not yet have published data.

The provision of nutrition information, as with the provision of any other form of information, is an enabling mechanism which allows the purchaser to make an informed choice about the product he or she is considering buying. The degree of interest in this particular segment of the mass of information printed on a food label is a matter that will be discussed in greater detail in section 8.4, but there are several issues that a manufacturer will need to consider in deciding whether or not to include nutrition information on the label:

- Is it required by legislation/what are the regulatory requirements?
- Can I conform with these?
- Will it be helpful to my customers/do they require this information?
- Are my competitors providing nutrition information?
- Will it give me an advantage over my competitors to provide nutrition information?
- Is there space on the label?

There may be trade-offs and these will need to be weighed in the balance, but two essential points should be borne in mind:

1. The provision of nutrition information on label is voluntary, *unless a claim is made*.
2. Approximately 80% of prepacked foods manufactured in the UK carry nutrition labelling, so not to do so is the exception rather than the rule.

The strict and detailed regulatory requirements which govern the presentation of nutrition information are set out in the next section, and it will be clear from a provisional glance that providing this information inevitably has cost implications, at least at the outset, unless the product is very simple and the values can be taken from a published source, such as McCance and Widdowson.[1]

The formulaic nature of the required presentation may appear unnecessarily prescriptive, but there is good reason for this. The underlying purpose of the current legislation was to encourage the provision of nutrition information in a prescribed format which would allow consumers to compare the nutritional content of various products. The effectiveness of this policy is another issue that will be discussed in greater detail in a later section, but it is not arbitrary and reflects the complexity of negotiating legislation on a contentious issue to cover a large trading block made up of more than a dozen European member states with diverse geographic and cultural backgrounds, and thus differences in local food supply and eating patterns.

Of the many factors governing food choice, of which price is likely to be quite high on most people's lists, nutrition information may not figure strongly for many. But the enormous number and variety of food products available on the market nowadays, including imports of exotic foods and ingredients from all over the world, resulting from the increasing interest in ethnic dishes generated by long-haul travel and TV cooks, not to mention new ranges of products inspired by these developments, means that the consumer needs ever greater knowledge and information to allow him or her to choose from this vast range. At point of purchase it is the food label that provides the information that will enable the consumer to make the choice between products. If diet and health are important to the consumer, the provision of nutrition information on the pack may be a deciding factor between purchasing the product and leaving it on the shelf, or, alternatively, a more careful study of the nutrition panel later in the home may influence a repeat purchase.

A further influence on the provision (or not) of nutrition information may be the intermediate customer, namely the retailer, rather than the end consumer. The major UK supermarket chains exert an enormous influence on the highly competitive retail market for food and thus on food production. All major retailers stock a wide range of 'own label' products, manufactured to their own specification by a variety of food manufacturers. The specification will cover not only the composition of the product but also the details of the food label. This will almost certainly include 'full' nutrition labelling, i.e. the 'Big 4' and 'Little 4' nutrients (see next section) and possibly additional, supplementary voluntary information, which is discussed in detail in sections 8.4 and 8.5. Manufacturers will need to consider whether or not they are producing supermarket 'own label' products and the competitiveness of their own branded goods if they are selling alongside them. Most retailers also carry a range of 'healthy' products, in which the composition is strictly formulated to meet specified nutrition criteria such as reduced fat content, lower sodium content, lower calorie, high fibre or any combination of these. They will invariably carry 'full' nutrition information. Consumers interested in a healthier diet may well look unfavourably at branded products that appear similar in content but do not offer nutrition information. Even if they are not specifically interested in this information they may wonder 'Have they got something to hide?'. In the UK, this has become a general accusation of consumer groups against those manufacturers who choose, for whatever reason, not to provide nutrition information.

In addition to retailer pressure and consumer demands, there is the added factor of government policy. Governments are the legislators and therefore have the power to regulate if they perceive a need or cannot achieve their aims by other means. Successive UK governments have placed diet and its influence on health under sharp focus in an effort to reduce the incidence of non-communicable disease, with the accent on prevention and the ultimate aim being to reduce the pressure and cost on the National Health Service of treating avoidable conditions. The factors influencing such diseases are recognised as multifarious and complex, but food is an easy target and food labels a visible, and therefore quantifiable, marker of change. Nutrition information is therefore an area of considerable government interest and is kept constantly under review.

The majority of food manufacturers support the provision of information which helps consumers make an informed choice about the products presented to them. This includes the provision of nutrition information. For as long as a voluntary regime continues, the decision as to whether or not to provide nutrition information is largely a commercial one, assessed against some of the factors mentioned above, and others which will be touched on later. The arguments for change will be discussed in detail later, but first let us consider in detail the requirements of the existing legislation.

8.2 EU nutrition legislation

8.2.1 Background

Legislation on nutrition labelling was developed as a means of providing consumers with information about the nutrient content of the foods they were choosing in a standardised format recognisable across the European Community, thereby also promoting the freedom of movement of goods in the Single Market.

An essential part of the philosophy behind the Nutrition Labelling Directive, the principle EU legislative instrument in this area,[2] was the growing recognition of the link between diet and health and the need to encourage consumers to make an informed choice about the foods they consume. It was considered that knowledge of the basic principles of nutrition and the provision of nutrition labelling would contribute significantly in this and act as a tool in the nutrition education of the public. To this end, it was deemed that the information provided should be simple and easily understood, with a standardised format which would allow comparison of one product with another.

Thus the dual principles underlying EU legislation on nutrition labelling are consumer information and education, and the removal of technical barriers to trade.

As usual in the development of harmonised legislation, one of the driving forces was the divergence in national legislation which risked causing reciprocal barriers to trade after completion of the Single Market on 31 December 1992. In the UK there was no specific legislation on nutrition information, but the Food Advisory Committee (FAC), whose advice carries considerable weight, had issued guidelines on nutrition labelling, which had been widely adopted by the industry. The Ministry of Agriculture, Fisheries and Food (MAFF) indicated its intention to introduce national legislation on the compulsory indication of fat content. This spurred the Commission into pushing its own proposals forward, on the basis that the UK's freedom to legislate in this area was constrained by its Community obligations under the Food Labelling Directive, 79/112/EEC (consolidated and updated as 2000/13).

Work on European harmonisation began in 1988, when the Commission put forward two linked proposals, one on compulsory nutrition labelling and the other setting out what that labelling should be. The directive eventually adopted in 1990, Directive 90/496/EEC on nutrition labelling for foodstuffs, did not require compulsory labelling, except where a claim is made, and focused more on the nature and format of the labelling, about which it goes into great detail.

Interestingly, for a piece of legislation for which one of the primary aims is the provision of information regarded as being of benefit to the consumer, it is a highly technical directive, unlikely to be understood by anyone without some knowledge of food science or food legislation, and ideally both. To understand and use it requires detailed analysis. Its main provisions are detailed in section 8.2.2.

8.2.2 Provisions of the current legislation

Format

The provision of nutrition labelling is voluntary, unless a nutrition claim is made, e.g. 'reduced fat', 'high fibre', 'low sodium'. If nutrition information is given, it must be in one of two formats:

either Group 1 information: energy, protein, carbohydrate and fat (in that order)

or Group 2 information: energy, protein, carbohydrate, sugars, fat, saturates, fibre and sodium (in that order).

These formats are commonly referred to as 'The Big 4' and 'The Big 4 plus little 4'. Quantities must be given per 100 g or 100 ml of the food or drink, or per 100 g/100 ml *and* per serving. The directive requires that the information be given in one place, in tabular format, with the numbers aligned if space permits.

Declarations may also be made in respect of vitamins and minerals, provided they are listed in the annex of the directive and are present in 'significant amounts', currently defined as 15% of the Recommended Daily Amount (RDA), supplied per 100 g or 100 ml of the food, or per package if the package contains only a single portion. The vitamins and minerals currently listed and their RDAs are given in Table 8.1.

A declaration may also be given in respect of one or more of the following:

- starch
- polyols

Table 8.1

Vitamin/mineral	RDA	Vitamin/mineral	RDA
Vitamin A	800 μg	Vitamin B_{12}	1 μg
Vitamin D	5 μg	Biotin	0.15 mg
Vitamin E	10 mg	Pantothenic acid	6 mg
Vitamin C	60 mg	Calcium	800 mg
Thiamin	1.4 mg	Phosphorus	800 mg
Riboflavin	1.6 mg	Iron	14 mg
Niacin	18 mg	Magnesium	300 mg
Vitamin B_6	2 mg	Zinc	15 mg
Folacin	200 μg	Iodine	150 μg

- mono-unsaturates
- polyunsaturates
- cholesterol

but if a declaration is made in respect of polyunsaturates, mono-unsaturates or cholesterol, the amount of saturates must also be given.

Calculation of energy value
For the purpose of calculating the energy value for these nutrients, the directive specifies the values on which they are to be calculated by means of the following conversion factors:

• Carbohydrate (except polyols)	4 kcal/g	17 kJ/g
• Polyols	2.4 kcal/g	10 kJ/g
• Protein	4 kcal/g	17 kJ/g
• Fat	9 kcal/g	37 kJ/g
• Alcohol (ethanol)	7 kcal/g	29 kJ/g
• Organic acid	3 kcal/g	13 kJ/g

Definitions
The directive, like most pieces of legislation, has to define what it refers to so all the nutrients that a manufacturer might want to put a number beside are defined scientifically. So, for example, the directive states that: '"protein" means the protein content calculated by using the formula: protein = total Kjeldahl nitrogen × 6.25', and '"saturates" means fatty acids without double bond'. Hence the earlier comment that it is a directive written for the food scientist rather than the average consumer.

The directive also defines 'average value'. This is necessary because the composition of foods is subject to natural variation due, for example, to variations in cultivar, weather, growing location, conditions and practices (for crops) and in breed, seasonality, rearing conditions and practices (for animal-derived materials). The directive therefore states that: '"average value" means the value which best represents the amount of the nutrient which a given food contains, and reflects allowances for seasonal variability, patterns of consumption and other factors which may cause the actual value to vary'.

Declared values
These are the average values of the nutrients, as defined above, based on:

- the manufacturer's analysis of the food;
- a calculation from the known or actual average values of the ingredients used;
- a calculation from generally established and accepted data.

The amounts declared must be for the food as sold. However, where appropriate they may relate to the foodstuff after preparation, provided that sufficiently detailed instructions for preparation are given and the information relates to the food as prepared for consumption.

The directive provides for the use of the Standing Committee procedure in the event of discrepancies being found between the declared values and those established during the course of official analysis. The Standing Committee is convened from experts from all member states who will adjudicate on the matter(s) placed before them.

In the United Kingdom, the term 'typical' is preferred to 'average' and is more generally used being a more representative indication of value than the average.

Nutrition claims

As stated earlier, the provision of nutrition information is voluntary unless a claim is made. So, for example, if a claim is made that a product is 'low in fat', at least Group 1 information must be given. Very often the full Group 2 information is given, but this would only be compulsory if the claim were for one of the 'Little 4' nutrients, so 'saturated fat' rather than 'fat'.

The directive defines as a nutrition claim:

> any representation and any advertising message which states, suggests or implies that a foodstuff has particular nutrition properties due to the energy (calorific value) it
> - provides,
> - provides at a reduced or increased rate or
> - does not provide
>
> and/or due to the nutrients it
> - contains,
> - contains in reduced or increased proportions or
> - does not contain.

Only generic advertising is excluded from this, so if a producer decided to launch a campaign to persuade people to eat more fresh green vegetables and claimed that green vegetables are low in fat, he would not have to include the nutrition information alongside his images of leafy greens.

Timescale

The directive entered into force in September 1990 and required that trade in products complying with the directive be permitted by 1 April 1992 and that products not complying with the directive be prohibited with effect from 1 October 1993.

The directive also required that, eight years from its notification, the Commission would submit to the European Parliament and the Council a report on the application of the directive and any appropriate proposals for amendment. This review, due in autumn 1998, has not yet taken place. This is discussed further in section 8.6.

8.2.3 Implementation

Most legislation is only as good as its implementation and enforcement. This has been mixed in the case of the Nutrition Labelling Directive. Some member states were tardy in implementing it into their national legislation and some, the UK being a prime example, did it so clumsily that it would have been a deterrent to use had the directive itself not already been familiar to most UK food and drink manufacturers and its provisions already widely used on a voluntary basis. Reports from elsewhere in Europe suggest that national implementing rules, which invariably entail a degree of interpretation, have indeed been a deterrent factor and have acted as a disincentive to provide nutrition information voluntarily. The UK's record of some 80% of manufactured food and drink products voluntarily carrying nutrition information remains a matter of surprise, admiration and consternation in other member states.

UK implementation of the Nutrition Labelling Directive is via the Food Labelling Regulations 1996 (as amended). These are complex regulations covering all the essentials of food labelling from batch marking to medicinal claims. Implementation of the Nutrition Labelling Directive, which took place in 1994, carried with it the usual burden of complexity that comes with turning the positive approach of EU legislation (you are not allowed to do it unless the directive says so) into the negative style of UK regulations (you can do what you like unless the regulations state that 'No person shall ...'). The transposition of Article 4.1 of Directive 90/496/EEC, which states simply that

> Where nutrition labelling is provided, the information to be given shall consist of either group 1 or group 2 in the following order:
> *Group 1*
> (a) energy value;
> (b) the amounts of protein, carbohydrate and fat.
> *Group 2*
> (a) energy value;
> (b) the amounts of protein, carbohydrate, sugars, fat, saturates, fibre and sodium.

became in Schedule 6A Part I of The Food Labelling (Amendment) Regulations 1994, a half-page single table listing both Group 1 and Group 2 nutrients, plus all the additional nutrients allowed to be mentioned, such as polyols under carbohydrates and polyunsaturates under fats, with a complex set of cross-references to Part II of the Schedule and subsequent paragraphs of Part I to explain the two separate groups and how they should be set out. No wonder MAFF needed to issue explanatory guidance notes to accompany the amendment to the Regulations.[3]

An additional complexity for manufacturers operating outside the UK has been differences in interpretation, and enforcement, of the legislation in other member states. This lack of uniformity in approach to implementation of EU legislation is by no means unique to nutrition labelling, but particular aspects of its implementation and enforcement in some member states appear to have acted as a disincentive to manufacturers to provide nutrition information. For example, for the first five years after adoption of the directive, a declaration of one of the 'Little 4', e.g. saturates or fibre, did not carry the obligation to provide full Group 2 information, whether or not the declaration of the nutrient was triggered by a claim, although after 6 October 1995 it did. Some member states never offered this distinction, whilst elsewhere the additional onus of having to provide all Group 2 information obliged some manufacturers to withdraw information on single 'Little 4' nutrients which they had previously provided.

Another complexity is that of unresolved issues such as the lack of an agreed EU definition of dietary fibre. The UK's adherence to the Englyst method of analysis and a definition of fibre as non-starch polysaccharides from plant cell walls only has hitherto left UK food manufacturers out of step with the rest of Europe (except Ireland), which favours the Association of Analytical Chemists' (AOAC International) method of analysis and includes lignin and resistant starch in the values for dietary fibre. While the United Kingdom has recently shown a willingness to take a pragmatic approach to this contentious issue by giving preferred status to the AOAC International method of analysis and setting a 'Guideline Daily Amount' for intake of fibre as measured by this method (as opposed to the Dietary Reference Value set by COMA),[4] an agreed European definition and recommended method of analysis is long overdue and urgently required. Article 1.4(j) of Directive 90/496/EEC states that '"fibre" means the material to be defined in accordance with the procedure laid down in Article 10 and measured by the method of analysis to be determined in accordance with that procedure'. Article 10 refers to the Standing Committee procedure described above. However, the matter was referred to the EU Scientific Committee for Food (SCF), an appropriate authority to pronounce on such an important matter. Unfortunately, the SCF has so far been unable to reach agreement on a definition of dietary fibre, which becomes ever more complex as nutrition science moves on. It is an unsatisfactory situation which the European Commission should seek to resolve.

8.3 Manufacturers' responsibilities

Clearly if a manufacturer decides to provide nutrition information, he must comply with the legislation, and if he wishes to make a nutrition claim on the product, he *must* give the nutrition information for that nutrient and all the others within the grouping. The above should provide a useful guide as to what is required, but it is important to check the precise requirements of the legislation in the country, or countries, in which the product is to be marketed. In the UK, the relevant legislation is the (now consolidated) Food Labelling Regulations 1996 (as amended). If marketing elsewhere in Europe, it is advisable to check the detail of the implementing legislation in each member state, though provided the provisions of the Nutrition Labelling Directive are complied with, the manufacturer is unlikely to fall foul of national implementation rules.

If providing nutrition information, it is important that it is accurate. Not only is it a legal requirement that any labelling information must be accurate and not misleading, but periodically consumer organisations run checks on the values given for the various nutrients and publicise embarrassing inaccuracies. Changes in values will inevitably occur with any changes in composition, so labelling information will need to be changed at the same time as the recipe. Whilst this may appear obvious, minor changes might easily be overlooked in terms of potential impact on nutrition information. It is important to establish the values accurately, preferably by periodical analysis, or by using a well-established database. Though expensive, recent analytical results would be indicative to any enforcement officer that the manufacturer had acted responsibly in the event of there being a dispute. Seasonal variation is provided for, but gross inaccuracy in terms of nutrient content and declared values is not, and reliance on published data may not suffice for more complex compositions. As in any other area of food legislation, a dialogue with the enforcement authorities, in the UK via the Home Authority,[5] is generally helpful, particularly for manufacturers who are unsure about their obligations, or about the precise requirements of the law. If, for example, the shape or size of the packaging precludes the inclusion of nutrition information in the recommended format, an opinion from the Home Authority on the acceptability of the alternative, proposed before the product is placed on the market, should reduce the potential for complaint afterwards.

The manufacturer also has an obligation to ensure that the label is understandable in the market(s) in which the product is sold. Fortunately this requirement has not yet extended to ensure that the consumer understands the nutrition information *per se*, only the language in which it is provided. This is important for products sold throughout the EU where multilingual packaging is used. Regrettably it cannot be assumed that consumers throughout the EU understand the nutrition information if given in the language of the country of manufacture, even though it is set out in a recognised format and order of nutrients. If other aspects of the label are translated into the language of the member state in which the product is being marketed, so must be the nutrition panel. This can be off-putting because of space, for which reason some manufacturers have, for many years, advocated a system of symbols for the nutrients which would be recognisable throughout Europe, and even internationally, but this has yet to be taken on board.

Finally, but importantly, the manufacturer has a responsibility to his customers. They may well express interest in the nutritional attributes of the product, whether or not nutrition information is provided. Many manufacturers and retailers produce leaflets to help explain nutrition labelling and how it can help them to choose a balanced diet. Manufacturers who do not produce their own leaflets can helpfully refer their customers to some of the organisations and resources referred to in section 8.7.

8.4 Consumer expectations

It is a long-held view of the UK food and drink manufacturing industry that nutrition labelling alone cannot educate the consumer to select a healthy balanced diet, but that it should provide the cornerstone of any nutrition education policy. Research has shown that relatively few consumers actually read the nutrition information provided, and even fewer of them understand it.

Nevertheless, consumer organisations clamour for more and more information, at least Group 2 nutrition labelling, and on a mandatory basis. The view of food manufacturers is that it is not always possible to meet all the expectations of consumers, either because they are not justified or because they are not feasible. For example package sizes vary. Cornflake packets provide ample space for all the requisite compulsory labelling information plus voluntary declarations, recipe suggestions, marketing offers and more besides. Individual chocolate bars and yoghurt pots do not. In addition, there are cost implications in the provision of nutrition information from analytical testing to packaging design and the manufacturer may not feel these additional costs are justified against the likely level of interest and consumer benefit in providing additional information.

In this context, research conducted by the UK Consumers' Association in 1995 revealed some interesting results.[6] A survey conducted in March/April 1995 questioned consumers on a number of issues about food purchases. The research was both qualitative and quantitative, the qualitative research consisting of four discussion groups held with women responsible for choosing and buying food. The quantitative research involved personal interviews with a representative sample of 1,454 people in Great Britain aged over 15 years – people responsible for choosing food and doing any of the food shopping. Respondents were asked which of certain attributes were important to them when shopping for food. The results are shown in Table 8.2.

Taking points 3, 6 and 8 as relevant to nutrition, diet and health, 23% of respondents consider this the most important aspect when shopping for foods,

Table 8.2 What shoppers look for[6]

	Most important N = 1454 (%)	Important at all (more than one answer) N = 1454 (%)
1. Price/value for money	34	87
2. Quality	21	77
3. Nutrition/how healthy it is	16	61
4. Family's/personal preference	12	53
5. How quick/easy to prepare	5	33
6. How fattening it is	4	33
7. Brand name/label	3	28
8. Special diet for medical reasons	3	10
9. Ethical/religious considerations	1	3

Table 8.3 Results of questionnaire where respondents were asked about their level of agreement with statements about information given on food labels (N = 1,454)[6]

		Agree (%)	Neither (%)	Disagree (%)
1.	I believe all of the health messages and claims that appear on products nowadays	33	17	48
2.	I never believe the nutrition claims on food packaging and always check the label for more information	47	24	27
3.	I find it difficult to work out from the nutrition information panel on food products what is good for me and what is not	49	18	32
4.	Nutrition information panels should be laid out in the same way for all food products	90	7	3
5.	Claims are a quick and easy way of seeing how healthy a food product is	54	17	25
6.	Health messages and nutrition claims should all be regulated	84	10	3
7.	It is important that nutrition information is shown on all food products	88	8	3

and a much higher number of some importance, even though nutrition and health ranked third after price/value for money and quality; 86% of those asked recognised a nutrition information panel, though of these only 42% took notice of it, with 33% stating that it was what they took most notice of.

Clearly, nutrition panels are very familiar and the circumstances in which nutrition information panels are used are of note:

- 36% when buying food not bought very often or never bought before;
- 34% when comparing two different makes or types of the same product;
- 26% when checking the nutrition claims made on the front of the pack;
- 15% never use this information;
- 15% every time food is bought that has this information on it;
- 14% have never seen this information.

The statements shown in Table 8.3 make interesting reading both for nutritionists and marketing departments. For the purposes of this chapter we shall focus on the third, fourth and final points and the preferences expressed by those who participated in the survey for presentation of the nutrition panel and the aspects and terms of the current format they found difficult to understand: the easiest to use were those that were clearer/easiest to read (e.g. large print/good for poor eyesight – 50%); good layout (general) (e.g. simple, clearer, neater, ordered (in a column), etc. – 22%); easy to understand (9%). Other points mentioned were highlighting/bold print; distinguishes between medium and high; shows value per 100 g; familiar/used to it/seen most often. The most difficult to use were those with poor layout (e.g. crammed together, jumbled, a muddle, words run together, cluttered, etc. – 41%); difficult to read/indistinct/

Table 8.4 Nutrition information: consumer confusion[6]

% RDA	80%
kJ	53%
kcals	41% calories (2%)
Sodium	14%
Saturates	11%
Per 100 grams	8%
Per serving	7%

small print (34%); not tabulated/itemised/in columns (9%); hard to pick out a particular piece of information (4%).

Table 8.4 lists a number of terms on a nutrition information panel that were found to be confusing.

Leaving aside the 88% desire for nutrition information on all food products against the 42% who actually take any notice of it, the provision of nutrition information is evidently the expectation and the norm. The next hurdle is making it helpful and meaningful to consumers.[6]

Alongside the results of the UK Consumers' Association's research, we shall also consider the Nutrition Labelling Study Report prepared for the UK Ministry of Agriculture, Fisheries and Foods (MAFF) by Research Services Ltd and published in April 1995.[7] The purpose of this study was to conduct a quantitative survey on consumers' awareness of nutrition labelling on food products, with the main objective of finding out the extent to which consumers use and understand this type of information. This was particularly timely in terms of baseline information as the new regulations on nutrition labelling, i.e. implementation of the Nutrition Labelling Directive, entered into force in March 1995. The study looked at the level of use of nutrition labelling; any problems in its presentation; and dietary habits, including changes in dietary patterns and what consumers thought they should be eating more or less of.

The areas examined were:

- The level of usage of nutrition labelling, including level of awareness; who uses nutrition labelling; the nutrients respondents were aware of/looked for on the label; use of 100 g or per serving information and which was most useful; how the labelling was used, i.e. to compare different foods or to assess the nutritional profile of individual foods and whether this was in the context of an individual meal or the whole diet.
- Any problems with the way in which the nutrition information was presented, including whether or not the nutrients shown were understood, i.e. energy, sodium, protein, etc., and the relationship between carbohydrate and sugar, fat and saturates; whether the units and other terms were understood, i.e. kilojoules, kilocalories, calories, grams, percent RDA; could consumers understand the relationship between per 100 g and per portion information? Were consumers able to compare nutrient levels in different products on a 100 g and per portion basis? And were consumers able to make judgements

about products across the range of nutrients, i.e. products that may be high in fat but differ in their content of saturates?
- The type of diet followed at the moment, including special dietary requirements; changes in diet over the last few years; what people think they should be eating more or less of.

A nationally representative sample of 1,000 interviews was used, following an initial qualitative phase for which actual knowledge was tested in a hall test situation. Those interviewed had to be personally responsible for shopping for food for the household or play a significant part in choosing what food was to be bought for themselves or their household. Further tests were incorporated to distinguish between those who were 'nutritionally aware' and those who were less well informed.

Most respondents were aware of nutrition labelling when asked about the information that could be found on foods, 62% mentioning nutrients whilst only 45% mentioned ingredients. The sample was more likely to look for nutrition labelling than look for the ingredients. Nutrition labelling was found to be the primary source of information about the content of foods. Around half the sample claimed to take this information into account when buying or using foods. People who were health conscious were more likely to take nutrition labelling into account than any other subgroup.

Within the nutrition panel, the information of most interest was fat levels (68%). About half those who looked at labels looked for energy levels, whilst sugar, protein and fibre were of less interest.

Overall, 'per serving' information was preferred to 'per 100 g' information (65% of those who looked at labels preferred per serving information against 21% preferring per 100 g information). However, the perceived usefulness of the 'per 100 g' information increased among respondents who could understand labels and had a high knowledge of nutrition.

Respondents were most likely to use nutrition information to compare two brands of the same product (49% of those who looked at labels claimed this). They were less likely to use it to compare two different products (only 15% claimed this). They were slightly more likely to use the information to assess products in terms of their whole diet rather than see how products fitted in with the rest of the meal.

The sample was equally divided between those who considered nutrition information useful and those who did not. However, certain subgroups believed it to be more helpful than others. A large majority (68%) of those who were health conscious considered the information useful. The researchers stated 'if it is the aim of nutrition labelling to be helpful to those who *want to use it*, it would make sense to see this group as the "target group", therefore a result of over two-thirds finding nutrition information useful seems encouraging.' Some 53% of females compared to 44% of males claimed nutrition information to be quite useful or very useful. Younger age groups were more likely to find the information useful (62% of 16 to 34 year olds, 51% of 35 to 44 year olds, 48% of

45 to 64 year olds and 33% of those aged 65+), as were those from higher social grades (60% of ABs, 53% of C1s, 48% of C2s, 43% of DEs). Other subgroups which found the information more useful were those with children in the household (57%); those whose education finished at the age of 19 or over (65%); those who were working (55%); those who were in the household with someone who had a special diet (60%); and those with high nutritional knowledge (61%).

Suggestions for improvement, apart from 'make it easier to understand', were generally in terms of making the print larger and giving more explanation of what the names and numbers meant.

With regard to understanding of the nutrients, fat was the most widely recognised nutrient and the one which respondents considered they knew most about; 89% claimed to know something about fat, and the vast majority stated that fat should be cut down on. When comparing two products, over half the respondents cited a lower level of fat as a healthier difference. By comparison, saturates, polyunsaturates and monounsaturates were less well recognised or understood. A proportion of the sample was aware of the need to cut down on saturates and increase polyunsaturates intake. However, compared with the number of respondents looking at fat levels on labels, very few claimed to look for saturate levels. When comparing two products, a very small number cited a lower level of saturates being a healthier difference. The researchers concluded 'it appears that there is a need for people to be made more aware of the implications of high saturate intake'.

Carbohydrate was very widely recognised, but around 25% of the sample claimed not to be sure what it was. Some 27% of respondents claimed carbohydrate intake should be increased whilst 17% believed intake should be reduced. When comparing two products, opinion was divided as to whether carbohydrate should be looked for in higher or lower levels. Fat, sugars, protein, fibre and energy were all more likely to be looked for on a nutrition label than carbohydrate. Sugar and starch were at least as well recognised as carbohydrate, sugar being understood and looked for more than carbohydrate itself. The vast majority of the sample believed sugar levels should be reduced. However, respondents were also far more likely to believe that starch intake should be reduced. The researchers concluded 'there is a need to educate people on the healthiness of this nutrient'.

Protein was widely recognised, but as with carbohydrate, about a quarter of the sample were not sure what it was. The majority of respondents agreed that intake of protein was beneficial and, when comparing two products, about half of them cited a higher level of protein as a healthier difference.

Fibre was better recognised and understood than the 'Big 4' nutrients, protein and carbohydrate. Some 84% of the sample claimed to know something about fibre. Over two-thirds of the sample believed fibre intake should be increased, and around half of the sample cited a higher level of fibre to be a healthier difference when comparing two products.

Energy was perceived by respondents as 'calories'. Most respondents (67%) claimed to know something about energy although, again, about a quarter of the sample were not sure what it was. After fat, it was the item of most interest on the nutrition label.

The term 'sodium' was much less well understood than 'salt'. Some 62% of respondents felt salt intake should be reduced, compared with 22% believing that sodium should be reduced. Sodium levels were rarely inspected by those who looked at nutrition labels. A majority of the sample mentioned a lower level of sodium being beneficial when there was a substantial difference in sodium levels between two products. Only 15% of the sample gave the same answers when comparing salt and sodium levels between two products, suggesting that relatively few respondents were able to equate sodium levels with salt levels.

'Calories' was a term recognised by almost all the sample, and most respondents correctly defined it as a measure of energy. By comparison, the terms 'kilocalories' and 'kilojoules' were less well recognised and understood. Grams were recognised and correctly defined as a measure of weight by the vast majority of respondents. Few claimed to be aware of the term 'percentage RDA', only about one in ten being able to define this term correctly.

As far as visualising what amount of food constitutes 100 g, the sample's performance was generally quite poor, though this was dependent upon the foodstuff in question. Where respondents were given three different amounts of the food to choose from, 28% of the sample gave the correct answer for fish fingers, 30% gave the correct answer for digestive biscuits and 16% gave the correct answer for raisins.

When asked to read figures from a nutrition label, or make comparisons of nutrient levels between two labels, almost a third of respondents were unable to answer each time. When no calculation was required, the majority of the sample could read 'per 100 g' or 'per packet' information from a label. The declaration 'of which saturates' or 'of which sugars' was understood as well as any other part of the label. Calorific information confused some respondents, bringing the proportion of those who could read this particular information down to around 50%. If a simple calculation was required, less than half of the sample were able to obtain the correct answer. This also applied when comparing nutrient levels between two labels. Most people did not have the ability to make the necessary calculations in their head to convert 'per 100 g' information into information for the whole packet, either when comparing two products or when assessing one product.

The researchers wanted to test how well respondents could assess the product's healthiness taking into account *all* as opposed to *individual* nutrients. In comparisons between two products, respondents were good at recognising the healthier product when the healthier differences between the products were to do with the most well known nutrients, for example lower in fat, higher in fibre, higher in protein. Respondents were less likely to recognise the healthier product when the healthier differences between the products involved nutrients which were less well known, for example, lower in saturates, lower in sodium.

Few respondents claimed to be in a household with someone who had special dietary requirements, the most common being a slimming diet which was mentioned by one in ten respondents. However, most individuals claimed to have changed their diet over the last few years, reasons to do with being healthier being the most commonly given. The foods people were most likely to believe they should be eating less of were fatty foods and sweet things. In terms of what the sample felt they should be eating more of, the most likely responses were fruit and vegetables. The researchers concluded 'most respondents claimed to be concerned about the healthiness of foods although less than half claimed to always choose a healthier food. A strong feeling for enjoying the foods they were eating emerged.'

Geographically, respondents in Wales and southern England performed better than those in the Midlands and the north of England. Respondents from Scotland and Northern Ireland were relatively less well informed.

The results of this research galvanised the food industry, both retail and manufacturing, into seeking ways of providing nutrition information in a manner that would be more helpful to consumers.

Further research commissioned and conducted in the UK by the Institute of Grocery Distribution (IGD) in 1996 showed that of those consumers who use the nutrition information on the label, most focus on energy, and to a lesser extent fat, slightly at odds with the RSL Report. The research also indicated that many consumers have little knowledge of how much energy, in terms of kcal or kJ, they need per day and little idea of what guideline targets are with respect to fat. As indicated in the previously listed research results, few understood the meaning of the term 'saturates', especially in the given format where it is indicated under fats as 'of which ...', and the concept of kilojoules was not understood at all. Most respondents said that 'per serving' information was generally found to be more helpful than 'per 100 g/100 ml', though the latter was useful when making comparisons between products at point of sale.

As a result of this research, proposals were drawn up for highlighting calories and fat on the nutrition label, and the UK Ministry of Agriculture, Fisheries and Foods (MAFF) and Department of Health (DH) were consulted about Guideline Daily Amounts of calories and fat, and values were agreed for both men and women. Various formats were tested on consumers and, as a result, a scheme for supplementary voluntary nutrition labelling was launched in February 1998.[8] The details of this scheme are set out in section 8.5.

8.5 Voluntary codes

Since the advent of the European Nutrition Labelling Directive and an agreed regulatory basis for the provision of nutrition information, itself voluntary unless a claim is made, there has been little scope for voluntary codes, except a general recommendation to UK manufacturers to provide at least Group 1 information (Food and Drink Federation), and a general understanding that manufacturers and retailers should do as much as possible to assist consumers to understand and use the information on the pack by providing leaflets, customer helplines and other sources of assistance. This is part of any major food business.

Feedback from consumers informs businesses about what their customers want and expect and the results of the UK consumer research outlined in section 8.4 above came as no great surprise to the food industry. Some major UK retailers had already begun to highlight information about specific nutrients below the standard nutrition panel in response to dietary advice in the 'Health of the Nation' White Paper[9] recommendations to reduce consumption of fat, and especially saturated fat, and to reduce levels of obesity.

In pursuit of UK government strategy in respect of nutrition goals, a Nutrition Task Force (NTF) was established to consider a range of aspects which might assist in improving consumers' eating habits including, unsurprisingly, the use of nutrition information. A group of experts drawn from the Nutrition Task Force and Food Advisory Committee (NTF/FAC Working Party) commissioned the consumer research project, described above, which concluded that current nutrition information was not helpful to many consumers.

The industry began to look at the possibilities, within the constraints of the existing legislation, for providing additional voluntary nutrition labelling as a tool to help consumers choose a healthy diet. The initiative was formalised in May 1995 under the auspices of the Institute of Grocery Distribution (IGD), a research organisation which draws its membership from every stage of the food supply chain and has links with a number of consumer associations.

The existing scientific, consumer and company research was reviewed, including that described in section 8.4, and new research commissioned to identify a labelling format for food products which would provide consumers with information to enable them to gain an improved understanding of the amount of fat and energy they consume in their daily diets. The objective was that the labelling format should provide relevant information and the nutrition information be expressed in a format useful to the consumer: it should help them to understand and manage the type and balance of nutrients (fat and energy) they were consuming in their diet. The information should be clear and simple to understand, for which reason the study focused on three nutrients in order not to confuse consumers with overly complex or detailed information. The choice of nutrients, fat, saturates and calories, resulted from the identification by the NTF/FAC Working Party, endorsed by the FAC, that such a focus would be a significant step forward in providing supplementary nutrition labelling, and that simpler supplementary nutrition labels would assist more consumers in selecting healthy diets.

A two-step research programme was conducted. Step one was qualitative research (five focus groups) which explored consumer attitudes towards nutrition labelling and provided guidance for the design of the major quantitative research. Step 2, the quantitative research, covered 2,300 adult consumers in a nationally representative study to assess the performance of a number of nutrition labelling formats. The research was designed to assess consumers' ability to use the label, i.e. their performance, rather than their preference for, different labelling formats.

The main findings of the research were as follows:

- Current levels of nutrition understanding are low.
- Current nutrition information is too complicated, frustrating and often illegible.
- Fat and calories were the most monitored nutrients, followed by protein and sugar, with fibre and sodium stimulating little concern.
- Many people only monitor the nutritional quality of their food if they are dieting or ill.
- The terms carbohydrate, saturates and sodium were not understood.
- Kilojoules are perceived as irrelevant to adult consumers.
- Consumers' ability to assess accurately the calorie content in foods was quite good. However, their ability to assess accurately the fat content in foods was poor. Products were rated as high or low in fat with very few intermediate assessments.
- Nutrition information is read whilst in the supermarket.
- There was genuine support for guideline daily amounts. Consumers felt that this was new information which helped them to place nutrition information in context, making the whole label more valuable and useful.
- Consumers preferred the use of whole numbers to decimal places and could not understand the relevance of having the information expressed to a tenth of a gram.
- 'Per serving' information was preferred over 'per 100 g', although the 'per 100 g' information was used when comparing the nutrient content of similar products at the point of purchase.

The IGD's guidance on Voluntary Nutrition Labelling was formulated after thorough analysis and discussion of the research results. The concept of Guideline Daily Amounts (GDAs) was central to the labelling formats tested and the underlying purpose of the supplementary voluntary nutrition information, i.e. to assist consumers in choosing a healthier diet in line with 'Health of the Nation' recommendations. A number of companies were already promoting daily 'amounts' for fat and calories in company literature, and it clearly made sense to work to a common standard to avoid confusing consumers. The GDAs recommended by the IGD were agreed following discussion with MAFF and the DH. They are based on the predicted daily consumption of an average consumer eating a diet conforming to Committee on Medical Aspects of Food Policy (COMA) recommendations. They are not intended as targets to aim for, but guidance to assist consumers in their understanding of their daily consumption of calories, fat and saturates. The recommended Guideline Daily Amounts are given in Table 8.5. It was recognised that consumer understanding of saturates is low, but the GDA was provided for those companies that choose to offer this information.

There are four other recommendations. First, additional on-pack information needs to be provided. In line with the research results which indicated that consumers were most interested in fat and calorie content, and on a per serving basis, the IGD recommends that this information be illustrated independently of

Table 8.5

Each day	Women	Men
Calories	2,000	2,500
Fat	70 g	95 g
Saturates	20 g	30 g

the nutrition panel in a separate box. Table 8.6 shows an example of nutrition information, and Table 8.7 illustrates a per serving example.

Where this is not possible because of the pack size or layout, it is suggested that this information be highlighted in colour within the nutrition panel. The 'per serving' measures must be stated and be appropriate to consumers, who show a preference for household units, e.g. per teaspoon, per half pack, per biscuit, per slice.

The next recommendation is that the column order in the nutrition information panel should be changed so that 'per serving' information comes before the 'per 100 g' information.

The IGD also recommended that a consumer education programme is required to improve consumer understanding about saturates and their role in the diet. Companies are recommended to use Group 2 nutrition information thereby ensuring that saturates appear on the label.

The final recommendation concerns legibility. The IGD referred to its June 1994 publication 'Packaging Legibility – Recommendations for Improvements' as guidance to assist consumers to read the information provided.

Table 8.6 Nutrition information

Typical values	per 100 ml
	280 KJ
Energy	67 kcal
Protein	3.2 g
Carbohydrates	4.8 g
of which Sugars	4.8 g
Fat	3.9 g
of which Saturates	2.6 g
Fibre	0.0 g
Sodium	0.1 g

Table 8.7 Per serving

Per serving (A cup)	
67 calories	3.9 g fat

The scheme has been widely adopted on UK supermarket 'own label' products. Uptake on branded foods has been less enthusiastic, for two reasons. First, the recommended supplementary format is, strictly speaking, illegal. This was recognised by the IGD, which stated in the published Guidelines:

> MAFF point out that, in the strictest interpretation of the current legislation, this information would likely fall within the definition of 'nutrition information' in Article 1(4)(a). Under Article 4, which sets out the order in which information should be given and the eighth Whereas clause, which prohibits any other form of nutrition labelling than that specified in the Directive, the presentation of fat and calories as recommended by the IGD Nutrition Group would be prohibited. However, LACOTS supports nutrition labelling which assists consumers to make informed dietary choices and takes the view that the IGD recommendations go some way to achieving this aim and therefore welcomes the IGD recommendations. Whilst noting the current legal constraints, local authorities will carefully consider pragmatic approaches which will benefit consumers. In the longer term LACOTS strongly supports changes to existing legislation to enable alternative forms of information to be given.

Second, most major food producers operate in a European environment and package and market accordingly. The supplementary information, especially the Guideline Daily Amounts, would not necessarily be appropriate to consumers elsewhere in Europe, and would almost certainly fall foul of local enforcement authorities. Many UK manufacturers have therefore opted not to display the supplementary information on the pack, but to include it in their company leaflets and promotional literature.

The IGD is committed to reviewing the effectiveness of the supplementary voluntary labelling, and to considering other nutrients. Sodium is currently under discussion.

8.6 Future trends

The application of the Nutrition Labelling Directive and its usefulness to consumers has remained under constant review since it entered into force. The anticipated arrival of the formal deadline of October 1998 for the European Commission to provide its report and any proposals for amendment provided an additional focus, as did the Commission's 1997 Green Paper on The General Principles of Food Law in the European Union. The review of the Directive is now well overdue and it is high time to consider proposals for change.

8.6.1 Proposals for changes to the legislation

In the light of experience and research into consumer use and understanding of nutrition information, the UK food and drink manufacturing industry's response to the need for change could be summed up in two words: simplification and flexibility. Most prepacked food and drink products sold in the UK already carry at least Group 1 information, and many provide Group 2 information. This reflects a genuine desire to take positive steps towards educating and informing consumers about the nutritional value of the products they consume. The Nutrition Labelling Directive is therefore perceived as a useful piece of legislation which could, if appropriate supporting education programmes were put in place, help consumers to construct a healthy balanced diet from the wide variety of products available to them. The legislation in its present form is not over-burdensome to industry, which sees no need for fundamental change. There is, however, scope for improvement in the light of experience, and indeed technological development, and some of the areas that might usefully be reviewed are as follows.

Format

Greater flexibility with Group 2 information would encourage more manufacturers to provide it. Where label space is at a premium, lists of nutrients with 'O' against them appear to waste it. So if the figure for the 'Big 4' nutrient is zero, the inclusion of the 'Little 4' nutrient appears superfluous, e.g.

Fat		0
–	of which saturates	0

Definitions

Technological advances necessitate a review of the definitions of several of the nutrients. For example the definition of carbohydrate encompasses substances such as polydextrose but, although a carbohydrate, polydextrose is only partially metabolisable and also demonstrates fibre-type properties. Moreover, the energy conversation factor of 4 kcal/g is far in excess of the acknowledged energy contribution of polydextrose: 1 kcal/g. Similarly fat replacers are now entering the market to meet the demand for lower energy foods. A substance such as olestra, already approved by the US Food and Drug Administration, is a lipid and, although not metabolisable, would be defined by the directive as a fat and attract an energy conversion factor of 9 kcal. This would clearly be a nonsense as the substance passes straight through the gut and provides no energy at all. Fibre has long been a bone of contention and despite years of discussion, the Commission has still failed to provide an agreed EU definition or method of analysis (see section 8.2.3), although with the UK recognition of AOAC method of analysis, this may be resolved in the not-too-distant future.

Simplification and flexibility

A concept enshrined in the directive is that nutrition information should be simple and easily understood. Highlighting specific information believed to be of most use to consumers, as recommended by the IGD, and removing any unnecessary clutter, would therefore appear to be a step in the right direction.

Many manufacturers would prefer to make nutrition information available on a 'per serving' basis with 100 g/100 ml as an option, rather than the other way round. As indicated above, this has already been shown to be preferred by consumers, who sometimes have difficulty in calculating the information for the amount of the product they would actually consume, especially if it is not a simple multiple or fraction of 100 g/100 ml.

Manufacturers would also like the flexibility to respond to consumer requests for additional information, which often occur on a short-term basis following media focus on a particular nutrient, such as the Willett study[10] which raised concerns about trans-fatty acids (TFAs). Currently there is no provision for such a declaration, though care would need to be taken not to encourage over-reaction to a 'scare' provoked by poor interpretation of a scientific report or a badly conducted study.

Manufacturers would also support the use of symbols to denote both macro- and micro-nutrients which would overcome any language problems and encourage more use of nutrition information on multilingual packs.

Simplification of the vitamin and mineral declarations would also be welcomed. The use of common synonyms is thought to be more helpful to consumers than the biochemical names prescribed in the directive, e.g. folic acid rather than folacin; and vitamin B1 instead of thiamin. The range of declarable micronutrients should also be extended to include important minerals such as selenium and chromium. It might also be helpful to be able to declare them in amounts other than 15% of the RDA per 100 g/ml or per serving, as a product might be an important source of a vitamin or mineral in terms of daily consumption but not meet the 'per serving' requirement, e.g. bread and milk.

8.6.2 Voluntary or mandatory?

The debate on whether nutrition labelling should be on a voluntary or mandatory basis has been going on since before the directive was adopted. Many consumer groups call for nutrition information to be mandatory on all prepacked foods and drinks. Arguably this should not present UK industry with any great difficulties as nutrition labelling is already provided voluntarily on about 80% of products, but what about the other 20%?

First, it should be remembered that no information comes without a price or a trade-off. Consider the amount of compulsory labelling on any food product: the name of the food; the supplier; a full list of ingredients; a use-by or 'best before' date; storage instructions; cooking or usage instructions to name but a few. Information overload can be off-putting. Consider then the cost of providing the nutrition information: working out the values for each product, monitoring and checking by analysis, formatting on the label and repeating this procedure wherever a change in the product is made. Even finding space on which to put the information may require new packaging design. Most people would agree that products that contribute a major source of the daily food intake might usefully provide nutrition information. But what about those products that are eaten only occasionally, possibly as a special treat, or which are used only in small amounts such as cake decorations, or only as an accompaniment, such as condiments? And should not more focus be placed on the general change in eating patterns and the tendency to consume more of our food outside the home, in restaurants and other catering establishments? At present the directive concerns nutrition labelling of 'foodstuffs to be delivered as such to the ultimate consumer'. It also applies to foodstuffs supplied to restaurants, hospitals, canteens and other similar mass caterers, but how often do we see nutrition declarations on a menu or on any food sold loose over the counter?

In any future trends, perhaps there is a need to reconsider the primary purpose of nutrition information and whether or not current practice is actually achieving it. Consumer information and consumer education are not necessarily the same thing. The primary purpose of the food label is to inform the consumer, not to educate, but the information, as indicated in section 8.4, is of little or no use without some pre-existing knowledge. Responsibility for providing this background knowledge has always been shared between government, consumer and health organisations, the media and the trade, but appears not yet to have fully achieved its aim. It would appear that the lead needs to come from government, in the United Kingdom specifically from the recently established Food Standards Agency, which has responsibility for food labelling and consumer information. If the provision of nutrition information is to assist consumers to choose a more healthy, balanced diet, they must first know what that diet should consist of, then how to use nutrition information to help achieve it. Consistency and simplicity in the messages would be a good start, followed by consistency and simplicity on the label. The growing use of electronic information, including in-store, offers opportunities not previously dreamt of. If such a medium can provide each individual consumer with every iota of information he or she wishes to know about any product, why try to cram more and more on the label? Perhaps future policy should gravitate towards providing only the more essential information on the label, and giving interested consumers quick and easy access to any other nutrition information they may wish to know via another medium. The debate on diet and health will continue indefinitely. Arguments over the provision of nutrition information will probably do likewise.

8.7 Sources of further information and advice

British Dietetic Association, 7th Floor, Elizabeth House, 22 Suffolk Street, Queensway, Birmingham B1 1LS. Telephone +44 (0)121 616 4900.

Campden and Chorleywood Food RA, Chipping Campden GL55 6LD. Telephone +44 (0)386 840319.

Food & Drink Federation, 6 Catherine Street, London WC2B 5JJ. Telephone +44 (0)20 7836 2460.

Institute of Grocery Distribution, Grange Lane, Letchmore Heath, Watford, Herts WD2 8DQ. Telephone +44 (0)1923 857141.
Leatherhead Food RA, Randalls Road, Leatherhead, KT22 7RY. Telephone +44 (0)372 376761.
Food Standards Agency, Aviation House, 125 Kingsway, London, WC2B 6NH. Telephone +44 (0) 20 7276 8000.

8.8 References

1. McCance and Widdowson's *The Composition of Foods*, Fifth Edition, The Royal Society of Chemistry and Ministry of Agriculture, Fisheries and Food, 1991.
2. *Official Journal of the European Communities*, No. L 276/40-44, 6 October 1990.
3. MAFF *Guidance Notes on Nutrition Labelling*, 1994, MAFF, London.
4. *Dietary Reference Values for Food Energy and Nutrients for the United Kingdom*, Report of the Panel on Dietary Reference Values of the Committee on Medical Aspects of Food Policy, HMSO, 1991.
5 The Home Authority Principle is defined by LACOTS, the Local Authorities Co-ordinating Body on Food and Trading Standards, as 'An authority where the relevant decision making base of an enterprise is located. It may be the place of the Head Office, or Factory, service centre or place of importation.' The purpose of this principle is to prevent infringements by offering advice at source and by encouraging enforcement authorities and enterprises to work in liaison with the Home Authority so as to encourage consultation and minimise duplication of effort and inconsistency in approach. LACOTS is seeking to promote this principle throughout the European Union.
6. *Consumers' Association Nutrition Labelling Research, Brief Summary of Quantitative Survey, 1995* (tables reproduced by kind permission of the Consumers' Association).
7. Nutrition Labelling Study Report, Research Services Limited (RSL), *MAFF J1366*, April 1995.
8. *Voluntary Nutrition Labelling Guidelines to Benefit the Consumer – Supplementary to legislative nutrition labelling requirements*, Institute of Grocery Distribution, 1998.
9. *The Health of the Nation – a strategy for health*, HMSO, 1992.
10. Willett, WC, Stampfer, MJ, Manson, JE, Colditz, GA, Speizer, FE, Rosner, BA, Sampson, LA and Hennekens, CH (1993). Intake of *trans* fatty acids and risk of coronary heart disease among women. *Lancet* **341**, 581–585.

8.9 Further reading

MAFF Foodsense Leaflet 'Use Your Label: Making Sense of Nutrition Information' MAFF 1996.

Council Directive of 24 September 1990 on nutrition labelling for foodstuffs (90/496/EEC), *Official Journal of the European Communities* No. L 276/40-44, 6 October 1990.

Department of Health, Guidelines on Educational Materials concerned with Nutrition, HMSO, London 1996.

The Food Labelling Regulations 1996, Statutory Instrument 1996 No. 1499, HMSO, London.

9

Nutrition and health claims

P. Berry Ottaway, Consultant, Berry Ottaway and Associates Limited, Hereford

9.1 Introduction

With the increasing market for healthy foods, functional foods and food supplements, health claims have become an important marketing tool.

The European Union Directive on Food Labelling (79/112/EEC) prohibits the attribution to any foodstuff of the property of preventing, treating or curing a human disease, or any reference to such properties.[1] Human disease has been interpreted as any ailment, injury or adverse condition, whether of body or mind. Any claim, expressed or implied, that a product can prevent, treat or cure a human condition or disease is regarded as a medicinal claim and the product has to be treated as a medicine under the requirements of Directive 65/65/EEC.[2]

Claims for specific biological functions of recognised nutrients such as 'calcium aids in the development of strong bones and teeth' or 'vitamin B_6 is important for the maintenance of a healthy nervous system' are regarded as nutrition claims and are normally permitted under food law. Other accepted nutrition claims relate to the amount of nutrients or components in a food such as 'low energy', 'sugar free' or 'a rich source of protein'.

A third set of claims are those that state or imply that the consumption of a food, or the component of a food, has a specific health benefit or avoids a specific aspect detrimental to health. Whilst a health claim can be a nutrient function claim such as 'vitamin E protects the fat in body tissues from oxidation', there are other health claims such as those made for the benefits of probiotic products which are not nutrient specific. During the latter part of the 1990s, health claims have become a more important aspect of food marketing and development, particularly in the context of functional food concepts such as the antioxidant nutrients, probiotics and prebiotics. Legislation constantly lags behind innovation and this has been the European situation with regard to health claims for foods.

As early as 1980 the European Commission recognised that the area of food claims required harmonisation and circulated the first proposal for a directive. After considerable discussion, agreement could not be reached between the relatively few member states at that time and the proposal was dropped. Over ten years later, in 1990, the Commission attempted to revive the proposals and then in 1992 there was a working document as a proposal for legislation on food claims from DG XXIV, the Consumer Policy Service of the Commission.[3]

Between June 1992 and June 1995 there were a number of revisions to the proposal as agreement could not be reached by the member states on any of the major points. In mid-1995 it was announced within the Commission that it would no longer be working on a directive for food claims. This meant that the individual national regulations would continue to remain in force, resulting in a diversity of approaches across the EU. The Commission then announced that it would try to introduce food claims principles into existing EU legislative texts such as nutrition labelling and misleading advertising.

By early 1999, this approach had also not succeeded, and the Commission began making further noises about developing EU legislation on health claims. The issue has become even more important with the growth of the functional food market in Europe. In January 2000 the Commission announced its intention to amend the Directive on Food Labelling (79/112/EEC) to specify conditions under which 'functional claims' and 'nutritional claims' may be made. The target date for adoption of the amendments by the Commission was given as July 2001, and adoption by the European Council and Parliament in July 2002.

The vacuum caused by the collapse of the EU harmonisation initiative on claims has, to some extent, been filled by national attempts to regulate this area. This has resulted in codes of practice on health claims being developed and introduced in Sweden, Belgium, the Netherlands and the United Kingdom. In addition, there has been a voluntary agreement in Spain between the Ministry of Health and the Spanish Food and Drinks Federation.

In France, the Conseil national de l'alimentation (National Dietary Council) has produced a draft opinion and proposals on claims linking diet and health. This follows a number of meetings of a working group from the end of 1997. During these meetings evidence was considered from consumer associations, industry and scientists.

Whilst all six countries have taken slightly different approaches to obtain the same objectives, they are all agreed on the major points. There appears to be a general consensus that some provision should be made to allow health claims for foods. These would be in addition to the nutrient function claims (e.g. calcium for healthy bones and teeth) which are already permitted, but would still exclude therapeutic claims.

9.2 Nutrition and health claims in the EU

By the beginning of 1999, there had been considerable progress made within five of the member states. There had been a general acceptance over the previous two to three years that the developing areas of functional foods and dietary supplements would require a significant re-evaluation of previously adopted positions on health claims. In addition, it was being conceded that legislation developed in the 1960s and 1970s was becoming inappropriate in the light of recent scientific developments in nutrition. For example, the concept of free radical damage to body tissues and the effect of antioxidant nutrients only goes back to the late 1970s. When the Medicines Directive 65/65/EEC and the Food Labelling Directive 79/112/EEC were developed, one of the primary distinctions between foods and medicines was the claims for 'prevention, treatment or cure'. At that time the term 'prevention' was interpreted in a prophylactic context such as prevention against malaria. It is now believed that the intake of certain components from foods, such as carotenoids and flavonoids, are the body's protection against some forms of free radical damage. Under present legislation, the preventative effects of these substances cannot be claimed as such for the foods. Whilst progress towards a review and possible revision of the term 'prevention' is not as fast as the food industry would like, this area is being given consideration at both European Commission and member state level. In the interim, a number of countries have made progress on internal policies for health claims.

9.2.1 Belgium

Work on a Code of Conduct on Health Claims was undertaken in Belgium by Fevia, the food industry federation. This work was completed in early 1999.[4]

Within the scope of the code are all health claims regardless of their nature, form or intended target group. The general definition of health claims covers all claims made for foodstuffs within the context of health promotion and subdivides these claims into four types:

1. *Nutrient function claims* describing the role of a nutrient or nutrients in the body's normal physiological processes. These claims are to be based on recognised and generally accepted scientific knowledge.
2. *Health effect claims* which refer to a specific positive health effect of a food or one of its components on the human body, on a physiological function, or some other biological parameter. These claims can also apply to ingredients and non-nutritive substances.
3. *Healthy eating pattern claims* which refer to official recommendations made by recognised national and international organisations relating to healthy eating habits, nutritional and other similar recommendations.
4. *Disease risk reduction claims*. These are defined as those that make reference to the fact that the consumption of a food helps to reduce the risk of contracting a particular condition or disease.

The code further explains and defines the terms 'health effect claim' and 'disease risk reduction claim'. It states that a number of statements such as 'to take care of', 'to support', 'to conserve', 'to optimise', 'to stimulate' and 'to affect' when used within the context of the maintenance of good health would normally fall within the definition of a health effect claim. Similarly, terms such as 'to increase', 'to reduce', 'to improve' and 'to reinforce' would be considered to fall within the category if they are used in connection with functions that in themselves do not imply an illness.

Disease risk reduction claims can be made if the effect of the food, or the effect of the food's nutrients or other components, has already been established on the basis of recognised and accepted scientific knowledge, or can be proven to help reduce the risk of contracting a particular illness. Such a claim must not be about prevention. The relationship between the food and the condition or disease should not relate directly to the prevention of the illness but rather about extolling the importance of a healthy diet in maintaining better health. The important aspects to be taken into consideration when making the claim are that the condition or disease in question is not already present; that the origin of the illness is multi-factorial; that the claim does not guarantee that the food can prevent the illness and that the claim indicates that the food contributes to risk reduction. An example of a disease risk reduction claim given in the code is 'Product X can contribute to the reduction of the risk of cardiovascular disease'.

It is an essential requirement of the code that the health claim must be scientifically justifiable and there is a section in the document covering this aspect. All the data used to provide the scientific substantiation for the claims must be collected into a dossier which must be retained by the person responsible for making the claim. The only requirement is that the dossier must be complete and must be available for examination if requested by the food inspection service. Unlike most of the other codes on health claims that have been developed in Europe, the Belgian code does not call for an independent assessment of the scientific evidence.

The scientific justification for a disease risk reduction claim must indicate whether and how the claim is based on recognised and generally accepted scientific relationships between dietary factors and specific illnesses. It must also demonstrate how the claim relates to official recommendations and any public information initiatives on the subject.

The code goes into considerable detail as to how the claims can be made. For example, they must not be misleading in any way and must be consistent with the scientific evidence in the dossier. They must also be as precise as possible in the description of the claimed effect and not make any extrapolation which cannot be supported by the scientific evidence.

All the other legal controls apply to the claims but the code makes some concessions provided strict conditions are applied. These allow certain words or descriptions to be used to explain the particular health claim more precisely. These can include words, synonyms or components of words that have a medical connotation (e.g. hygiene, medical or disease) and the names of diseases or representations of a disease or of people suffering from a disease. The concession also allows the names or representations of organs, blood or the circulatory or nervous systems and the effects of the food on them.

All the above terms or statements must not be used in a way that indicates or implies that the food has a preventative, therapeutic or curative action. A health claim or the form of advertising or promotion in which it is included must not suggest or imply that good health is negatively affected by the non-use of the particular food. It must also not suggest or imply that the use of the food guarantees the maintenance of normal good health. The code contains a number of other similar requirements or restrictions in advertisements or promotional materials in which a claim is made. This includes a section on the use of testimonials which may be permitted under certain conditions.

9.2.2 France

During a plenary session held in September 1997, the French Conseil national de l'alimentation (CNA – National Dietary Council) undertook the task of considering claims linking diet and health, including functional claims.[5] The deliberations of the Council have resulted in an opinion following a wide consultation which included consumer associations, the food and drink federations and food scientists. The Council has also heard evidence from people involved in advertising and from market researchers studying dietary behaviour.

The opinion of the Council was that the prohibition on therapeutic claims must remain but, provided certain principles were maintained, health claims would be acceptable. Claims that could be permitted with suitable controls were:

- a claim for the food product as aiding a reduction in risk of a disease
- a claim for a positive contribution to health when presented as having an influence on the modification of a physiological state or biological parameter
- a functional nutritional claim that describes the positive role of the nutrient in the normal functioning of the body.

The Council felt that such claims would require prior authorisation before marketing unless they appeared on an official approved list.

For new claims, and possibly for all claims, the validation of the scientific data that form the basis of the claims should be undertaken by independent organisations such as the Human Nutrition Research Centres. The Council also proposed that a brief trial period should be introduced for the recommended system before it became permanent. The Council was concerned that the quality of the scientific justification should be of the highest standard and that there should be more emphasis on punishing those making fraudulent claims.

9.2.3 The Netherlands

After a considerable amount of discussion, a Code of Practice for assessing the scientific evidence for health benefits from food and the health claims made for such foods and drinks was drawn up on the initiative of the *Voedingscentrum* (the Netherlands Nutrition Centre). This code, which was completed in April 1998, contains a procedure for the assessment of the scientific evidence for health benefits stated in health claims.[6]

The code was developed under the auspices of the Dutch Food and Drink Industry Associations, the Consumers Associations, the Association of Dutch Advertisers, Retail Associations and the Netherlands Nutrition Centre. The code was presented to representatives of the Ministry of Public Health, Welfare and Sport, the Ministry of Agriculture, Nature Management and Fisheries, and the Dutch Association of Branded Product Manufacturers at a meeting on 21 April 1998. The government representatives welcomed the procedures laid down in the code and declared official support for it. However, this code of practice has not been officially adopted by the government.

Under Dutch law, a health claim is defined as 'A direct, indirect or implied claim that a food carries special qualities which improve or maintain the user's health'. This is interpreted as a claim made by any promotional means and by any means of communication. Products imported into the Netherlands have to meet the same criteria. The Dutch code of practice does not cover medical claims nor does it apply to nutrient content claims or product safety, as these aspects are already covered by law.

The code requires that health benefit claims are assessed by an independent panel of experts appointed by the Netherlands Nutrition Centre. This assessment covers the scientific content of the supporting evidence for the claim. There is a primary requirement that the evidence must be based on relevant data from human subjects. It must be demonstrated that any effectiveness of the substance or substances determined from research is not reduced when included in a commercial product as presented to the consumer. The data used must be relevant to normal use of the product by the target population and the health benefit claimed must also be relevant to the target population. It is also necessary to show that the scientific evidence is reproducible. The code also has a requirement that the claimed health benefit does not clash with any dietary guidelines laid down by any of the authoritative bodies in the country.

The expert panel appointed to assess an application is only required to give an opinion on whether the scientific evidence is sufficient to support the claim. The code specifically states that it is not up to the panel to assess whether the health benefit is correctly presented in the claim. This is given as being the responsibility of the Advertising Code Foundations, its Appeals Tribunal and Dutch courts of law.

The code has detailed requirements for the composition of the panel of experts and the way in which they have to operate. Assessments must be completed as quickly as possible but a decision must be given no later than three months after the official receipt of the dossier. If requested, the panel must give the applicant the opportunity to explain his case in person. If the panel fails to reach a unanimous decision, the application must be accepted on a majority vote. However, there is provision for the objecting members to state their reasons in the final report.

When an applicant objects to the panel's decision, the Nutrition Centre can be asked to convene a second panel of experts. Except under certain conditions outlined in the code, the Nutrition Centre has to respect confidentiality and the data and assessment report are not to be made available to third parties.

There is a requirement that the code of practice operation is reviewed on a regular basis, at least every two years, by the Nutrition Centre.

9.2.4 Spain

An agreement on health claims for foods was signed on 20 March 1998 between the *Ministerio de Sonidad y Consumo* (Spanish Ministry of Health) and the *Federación de Industrias de Alimentación y Bebidas* (FIAB – the Spanish Federation of Food and Drinks Manufacturers).[7] This agreement was voluntary and clarifies the situation relating to health claims within the legislation covering the labelling and advertising of foods. Within the agreement, health claims are defined as any claim relating to:

- the function of one or more nutrients in the human body
- the effects of the consumption of one or more food products on human health
- the consumption of certain foods as part of a healthy diet.

In common with the legislation of all other member states of the EU and as a consequence of Directive 79/112/EEC, Spanish law prohibits claims that attribute or imply preventative, curative or therapeutic properties to a food. The primary purpose of the agreement is to lay down the guidelines which permit health claims in the categories listed above to be made without infringing the law.

The agreement is intended to apply to the labelling and advertising of all foods and drinks with the exception of products which come under the classification of foods for particular nutritional uses (PARNUTS) and mineral waters. Nutrition content claims or claims such as 'contains vitamin A' or 'contains iron and calcium' are also excluded from the scope of the agreement as they are already controlled by specific legislation.

There is a general requirement that all health claims must be truthful and be able to be clearly substantiated by scientific evidence. Where a claim is made for the beneficial properties of an ingredient rather than the food itself, the nutrient or component that is claimed to have the beneficial properties must be present in the food in sufficient quantities to produce the claimed effect. The same applies to claims for the absence of a specific component or nutrient (e.g. saturated fat).

There is a requirement that whenever a health claim is made in the labelling or in promotional materials for a food it should be accompanied by a statement about the importance of maintaining a healthy and balanced diet. Health claims must also, where appropriate, comply with the rules laid down for the nutritional labelling of foods. It is not permitted to claim a particular beneficial effect for a specific brand when all foods in the same category have the same properties. As

Table 9.1 Spain: examples of unacceptable health claims (annex to agreement)

Avoids or replaces medication	Cures constipation
Helps fight osteoporosis	Boosts your immune system
Fibre prevents colon cancer	Enhances your natural defences
Tonic effect or action	Avoids or replaces medication
Relaxing effect or action	Product X helps you lose weight
Brand X oranges are recommended by WHO	

an annex to the agreement, there is a non-exhaustive list of claims which have been deemed to be unacceptable. Examples are given in Table 9.1.

The agreement requires that, as a general principle, all health claims must avoid categorical statements guaranteeing the beneficial effects claimed for the food. Instead, words such as 'aids', 'facilitates' and 'eases' should be used to indicate that the consumption of the food helps to achieve the beneficial effects but that other factors such as a healthy diet, lifestyle and physical exercise must also be taken into consideration.

As part of the agreement, a joint committee composed of officials from the ministry and representatives of the food industry has been set up to review proposed claims which can be submitted on a voluntary basis before marketing. The committee is required to give an opinion within ten days of submission. Failure to respond to an application within the prescribed time means a tacit favourable opinion on the use of the claim.

9.2.5 Sweden

Sweden was the first of the six EU countries developing guidelines on health claims to agree to adopt a self-regulating procedure. The first programme to regulate health claims was agreed by the food industry and came into effect in August 1990. This was monitored for three years until July 1993 and revised self regulating rules were agreed in August 1996 and came into effect from 1 January 1997.[8] This revised programme was developed and agreed by representatives of all sectors of the Swedish food industry including the Federation of Swedish Food Industries, the Swedish Food Retail Association, the Federation of Swedish Farmers and the Grocery Manufacturers of Sweden.

In the explanatory document, a health claim is defined as 'an assessment of the positive health effects of a foodstuff, i.e. a claim that the nutritional composition of the product can be connected with prophylactic effects or the reduced risk of a diet-related disease'. This definition is qualified by the requirement that the health claim must be based on the importance of the product in a balanced diet and must be in line with the official Swedish dietary recommendations. Table 9.2 gives diet-related risk factors which may form the basis of health claims in Sweden.

A health claim must consist of two parts. The first must provide information on the diet and the health relationship of the food. The second part is to consist

Table 9.2 Diet-related risk factors which may form the basis of health claims in Sweden

1. Obesity	5. Constipation
2. Cholesterol level in blood	6. Osteoporosis
3. Blood pressure	7. Dental caries
4. Atherosclerosis	8. Iron deficiency

of information on the composition of the product. Two examples given as acceptable health claims are:

- Part 1. Iron deficiency is common among women but can be prevented by good dietary habits.
  Part 2. Product X is an important source of the type of iron that is readily absorbed by the body.
- Part 1. Omega-3 fatty acids have a positive effect on blood lipid and can therefore help protect against cardiovascular disease.
  Part 2. Fish product X is rich in omega-3 fatty acids.

Six other examples of approved health claims are given in the document. These include claims for constipation and dietary fibre, osteoporosis and calcium rich foods, atherosclerosis and low saturated fatty acids and/or low salt content, and dental caries and sugar-free products. With each approved claim there is an example of an unapproved statement or reference regarding the product composition and effect. For example, it is permissible for the second part of a claim relating to cholesterol and cardiovascular disease to say 'Brand X contains a low amount of saturated fat and total fat' but it is not permissible to say 'Brand X provides excellent protection against cardiovascular disease through its low content of saturated fatty acids' or 'Brand X will help you reduce your blood cholesterol levels'.

Also included in an appendix are examples of approved nutrient function claims such as 'contains zinc which is a component in many of the body's enzyme systems'. Nutrient function claims can only be made for products that contain a significant amount of the nutrient in question. The term 'significant amount' is taken from the definition in the Nutrition Labelling Directive 90/496/EEC which is that the product should contain at least 15% of the recommended daily intake per 100 g or 100 ml of the product, or in a package containing one serving of the food.

Claims relating to the effect of a food on the body's blood sugar level after a meal may fall within the framework of nutrient function claims, provided that the claim is not related to disease or the risk of disease.

It is stated in the document that it is the intention of the industry to publish a supplement to the current rules to take into consideration the use of claims for functional foods. Manufacturers who wish to use a claim for the health promoting effects of a product which is not in the approved list can, provided the claim can be substantiated, notify the intention to use the claim to the National Food Administration or apply for registration as a food for particular nutritional use (PARNUTS). Experts from the Swedish Nutrition Foundation are available to advise companies on the objectivity of proposed health claims. In their review of the claims and supporting data the appointed experts are required to pay special attention to the interests and needs of the average consumer regarding guidance on the nutritional aspects of the food.

Provision has been made as part of the initiative for the organisations and companies to attend seminars in which the Swedish Nutrition Foundation will be involved. The seminars will be for those involved in the marketing of food to understand the relationship between diet and health and the legal issues involved in making claims, particularly health claims.

The organisations that are signatories to the rules are required continuously to provide relevant information on the above issues to companies in the food industry. Each organisation has appointed a contact who is responsible for maintaining communications with the authorities and the Swedish Nutrition Foundation and to disseminate the information to their member companies.

9.2.6 United Kingdom

In December 1996 the Food Advisory Committee (FAC) completed its review of the British market for functional foods and the control of health claims. As part of this review, the FAC published draft guidelines on health claims for foodstuffs.[9] These guidelines were intended to set out the conditions which food manufacturers or retailers were required to follow when making health claims for their products. The scope of these guidelines was that they applied to all foods and drinks, including food supplements, but with the exception of those products controlled under the EC Directive on foods for particular nutritional uses (PARNUTS).

A health claim is defined in the draft as any statement, suggestion or implication in food labelling or advertising that a food is beneficial to health. Nutrient content claims and medical claims are excluded from this definition. Health claims could be subdivided into three categories:

1. Claims that refer to possible disease risk factors (e.g. can help lower blood cholesterol).
2. Nutrient function claims (e.g. calcium is needed to build strong bones and teeth).
3. Recommended dietary practice (e.g. eat more oily fish for a healthy lifestyle).

The draft guidelines devoted a section to the principles underlying health claims and one to the scientific substantiation of claims.

Following a period of consultation, which in general met with a positive response, the Joint Health Claims Initiative (JHCI) was established in June 1997. This was a joint venture between consumer organisations, enforcement authorities and industry bodies with the objective of establishing a code of practice for the use of health claims for foods.

During the development of the code a strong consensus emerged between the various organisations involved that the existing legal and enforcement framework governing claims in the UK was both incomplete and inflexible. There were also a number of areas of uncertainty leading to subjective interpretation. There was general agreement amongst the participants that there was a need for a review of the laws on claims in the light of scientific advances. A recommendation was made to the ministers to re-assess the existing legislation and its interpretation.

The draft code of practice which was developed by the JHCI further separates health claims into generic claims and new claims.[10] Generic health claims are defined as those based on well-established and generally accepted knowledge, evidence in scientific literature and/or recommendations from national or international public health bodies. A regularly reviewed and updated list of generic health claims was to be maintained as part of the administration of the code. As part of the preparation for such a list, the Medical Research Council's Human Nutrition Research group was asked in 1999 to consider the scientific evidence substantiating fourteen putative generic claims. Their report, which was published in early 2000, could not find good evidence for some of the claims.[11]

A new health claim must be based on scientific evidence as applied to either existing or new foods. The scientific evidence must be substantiated in accordance with detailed rules laid down in the code. Companies wishing to market a food with a new health claim must show that the claim is likely to be true and that the evidence in its support outweighs any opposing evidence or opinions. Companies are also considered to be responsible for ensuring that the validation of a new health claim is carried out alongside all other checks necessary when assessing the suitability of the food for marketing.

There are a number of criteria that must be met when using a health claim. It is essential to demonstrate that the food, or its components, will cause or contribute to a significant and positive physiological benefit when consumed by the target population as part of their normal diet. The effect must be achieved by the consumption of a reasonable amount of food on a regular basis or by the food making a reasonable contribution to the diet. The claimed effect must be maintained over a reasonable period of time and it should not be a short-term response to which the body adjusts, unless the claim is relevant for only a short- or medium-term benefit. An example of the latter would be the use of folic acid just before and during the early stages of pregnancy in case of neural tube defects.

The substantiating evidence is required to demonstrate the minimum or maximum amount and the frequency of consumption of the food required to achieve the effect. Alternatively, it must be shown that the food can provide a reasonable dietary contribution to that required to achieve the effect. The substantiation should also explain how the effect is brought about. The code accepts that the exact biological mechanism need not be fully understood or explicable. The types of individuals who can most benefit from the effect must also be indicated, particularly as to whether the effect could apply to the whole population, population sub-groups, at-risk groups or individuals with a particular disease.

The scientific data needed to support a claim is described in some detail in the code. The claim must be based on a systematic review of all the available evidence, including published scientific literature relating to the validity of the health claim. The conclusions drawn from this review should be based on the totality of the evidence and not just on the data that support the claim.

The conclusions should be based on human studies which are the most methodologically sound available or other human evidence and not just on biochemical, cellular or animal studies. Whilst, ideally, the conclusions should be based on experimental studies in humans, it is accepted that observational studies may be acceptable in some circumstances. The research should assess the effect of foods on the health status of human subjects. Where obtaining full clinical evidence is difficult, lengthy or expensive, it would be acceptable under the code to provide evidence of the effect of the foods on bio-markers, provided there is a strong correlation between the bio-markers and a relevant indicator of well being, disease or risk factor.

Companies wishing to make a health claim should submit the details of the claim and the substantiating scientific evidence to the Code Administration Body (CAB). The CAB will seek the opinion of a panel of experts which is described in the code as the Expert Authority. The CAB will consist of a Council and a Secretariat. The Council is conceived as a tripartite body, with representation from enforcement, consumer and industry groups, and has the responsibility for monitoring the code, for the selection of members of the Expert Authority and for considering any amendments to the code in the light of experience. The Secretariat will service both the requirements of the Council and the Expert Authority and act as the primary point of contact for all interests. By the middle of 1999 the organisation and composition of the Expert Authority had not been resolved. It was anticipated that the code would come into effect in the middle of 2000, provided agreement was reached on the composition of the Expert Authority, and there was sufficient funding to support the Secretariat. The organisation formally started in December 2000.

9.3 Substantiation of health claims

Common to all the European national initiatives on health claims is the need for good scientific substantiation of the claim. All the initiatives require that the scientific evidence is of good quality and it is either expressed or implied that the science should sustain peer-review. The British code of practice details the type of data that should be presented. Both the British and the Dutch codes have a primary requirement that the evidence must be based on relevant data from humans. There is also a need to demonstrate that the food, when consumed in reasonable quantities, can achieve or contribute to the claimed beneficial effect. The Dutch code has a requirement that it must also be demonstrated that the effectiveness of any substance or substances as determined by research is not reduced when incorporated into a commercial product. When considered collectively it is probable that the requirements for scientific substantiation will have a limiting effect on the number of non-generic health claims that will receive approval.

9.4 Nutrition labelling

In most instances, a nutrient function claim or a nutrient related health claim will require nutrition labelling. The rules for nutrition labelling are laid down in Directive 90/496/EEC and cover the labelling of the macro-(energy) nutrients, cholesterol, sodium, twelve vitamins, six minerals and fibre.[12]

The factors for calculating energy differ from many countries outside the EU in that polyols must be calculated at 10 kJ and 2.4 kcal per g and organic acids at 13 kJ and 3 kcal per g. With some products these values can lead to significant differences between the United States and Europe in the amount of energy that can be claimed. There are also considerable differences between the United States and Europe in the calculations of the activity of some vitamins.

Since the directive was adopted in 1990 there has been a continuing debate on the definition and calculation of dietary fibre. No useful definition is given in the directive and this has led to differences in interpretation between member states.

9.5 The future

The development of functional foods in Europe has been increasing rapidly since 1997 with a number of multinational food companies taking an active part in the growth of this sector. In addition, the food supplements market is still showing growth, particularly in countries on mainland Europe. Paramount to the growth of both areas is the ability to make claims in the marketing of the products. As discussed, this has already been appreciated and addressed by six countries in the EU who have some form of agreement or code of practice acceptable to the industry, the enforcement authorities and consumer interests. The other countries in the EU, such as Germany, which have not already embarked on the establishment of procedures for handling health claims, are, in most cases, in early internal discussions on the subject.

It can be predicted that the inconsistencies between the existing codes will probably lead to barriers to trade in the products on a pan-European basis. For example, the Belgian code of practice requires only that a dossier containing the scientific evidence be retained for examination on request, whereas in the Netherlands and United Kingdom an independent review of the data will be the norm. There has been so much national activity in the area of health claims that the European Commission is forced to re-activate its work on the proposed directive. It may now be even more difficult to achieve harmonisation as a number of the member states will fight to retain their recently developed national positions.

A further, and more complex aspect, is that some products are being developed in such a way as to blur the border between foods and medicines. The Medicines Directive 65/65/EEC contains the definition of a medicine which is in two parts. The first part relates to products which are intended to prevent, cure or treat a human condition or disease. The prohibition of claims or implications of prevention, treatment or cure has been carried over into European, and hence national, food law.

What is more difficult to interpret is the second part of the definition which states: 'Any substance or combination of substances which may be administered to human beings or animals with a view to making a medicinal diagnosis or to restoring, correcting or modifying physiological functions in human beings or in animals is likewise considered a medicinal product.'

The appearance on the market of fat spreads (e.g. margarine) containing increased levels of phytosterol esters for the reduction of cholesterol, and products aimed at women containing high levels of added isoflavones, is making it more difficult to identify clearly the legal status of a product. In the past the 'rule of thumb' has been that products that maintain the healthy functions of the body's organs, tissues or systems are regarded as foods, whilst those that are developed or promoted to induce changes to the normal physiological functions of the body may be medicinal. Deliberately inducing diuresis, for example with a product designed for slimmers, could be considered to be modifying the normal functions of the body's renal system. In such a case the classification would pivot on the intent of the product, as diuresis can also be a side effect from the consumption of foods or drinks containing substances such as caffeine or alcohol.

The developments in the functional food area that are in progress may require either a more liberal interpretation of the definition of a medicine or a revised definition. Article 1 of Directive 65/65/EEC includes the term 'preventing disease'. Over thirty years ago when the directive was adopted, the concept of free radical damage at a cellular and tissue level did not appear in the medical or nutritional literature. More recently it is generally accepted that the consumption of the antioxidant nutrients can play a significant role in the retardation of the development of certain diseases. Under current European legislation it is illegal to either express or imply that a product rich in antioxidant nutrients can play a preventative role with these diseases.

The positive developments on health claims by some of the member states of the EU and the discussions on functional foods that have been initiated at European Commission and national government level should lead to a modernisation of the legislation but, unfortunately, not in the short term.

9.6 References

1. *European Council Directive 79/112/EEC* relating to the labelling, presentation and advertising of foodstuffs. *OJ* of EC L33/1 of 8 Feb. 1979.
2. *European Council Directive 65/65/EEC* relating to proprietary medicinal products. *OJ* of EC 22 Feb. 1965.
3. European Commission. Draft proposal for a Council directive on the use of claims relating to foodstuffs. *SPA/62 ORG*: Fr of 26 June 1992.
4. Federatie Voedingsindustrie/Fédération de l'Industrie Alimentaire (Belgium). *Health Claims Code of Conduct*, draft, 21 Oct. 1998.
5. Conseil national de l'alimentation (France). *Claims linking Diet and Health*. Draft opinion version No. 8, 1998.
6. Voedingscentrum (Netherlands), *Code of practice assessing the scientific evidence for health benefits stated in health claims on food and drink products*, April 1998.
7. Joint 'Ministerio de Sonidad y Consumo' (Ministry of Health, Spain) and Federación de Industrias de Alimentación y Bebidas (Spanish Federation of Food and Drink Manufacturers) *Agreement on health claims on foods* of 20 March 1998.
8. Federation of Swedish Food Industries et al. *Health Claims in the Labelling and Marketing of Food Products. Food Industry's Rules (Self-Regulating Programme)* Revised programme of 28 Aug. 1996.
9. Ministry of Agriculture, Fisheries and Food (United Kingdom), *Food Advisory Committee Review of Functional Foods and Health Claims*, 16 Dec. 1996.
10. Joint Health Claims Initiative (United Kingdom). *Code of Practice on Health Claims on Foods*. Final text, 9 Nov. 1998.
11. Medical Research Council (2000) 'Validation of generic health claims'. Briefing paper by MRC Human Nutrition Research, Cambridge, England.
12. *European Council Directive 90/496/EEC* on nutrition labelling for foodstuffs. *OJ* of EC L276/40 of 6 Oct. 1990.

10

Foods for particular nutritional uses (PARNUTS)

P. Berry Ottaway, Consultant, Berry Ottaway and Associates Limited, Hereford

10.1 Introduction

Foods for particular nutritional uses (which have assumed the acronym of PARNUTS) are foods which are designed to fulfil a particular nutritional function and are clearly distinguishable from foods for normal consumption. Good examples of these products are the infant formulae and weight control diet products. From the early days of the development of European food law this was a category of foods for which it was considered that specific controls would be required. The first directive on Foodstuffs for Particular Nutritional Uses was Directive 77/94/EEC of December 1976.

During the changes to European food law which resulted from the 1985 White Paper, it was decided that the PARNUTS category should be covered by a framework directive, and after considerable discussion a Common Position was agreed in 1988 and adopted as Directive 89/398/EEC of 3 May 1989.[1] This has subsequently been amended by Council Directive 96/84/EC[2] and European Parliament and Council Directive 99/41/EC.[3]

This framework directive covers all foods designed and marketed for particular nutritional uses. These are defined in the directive as:

> foodstuffs which, owing to their special composition or manufacturing processes, are clearly distinguishable from foodstuffs for normal consumption, which are suitable for their claimed nutritional purposes and which are marketed in such a way as to indicate such suitability.

To qualify as a particular nutritional use the foodstuff must be able to fulfil the specific nutritional requirements of certain categories of people whose digestive processes or metabolism are disturbed (e.g. diabetics or coeliacs) or people who are in a special physiological condition and who are therefore able to benefit from controlled consumption of certain substances in foodstuffs (e.g. people who are controlling their weight). The third category of products is very specific and covers foodstuffs for infants and young children in good health.

It is a prerequisite of the directive that the nature or composition of the product must be appropriate for the particular nutritional use for which it is intended and it is also required that the product complies with any mandatory provisions applicable to foodstuffs for normal consumption except for any changes necessary to ensure their conformity to the required use. This means that the products have to conform with all applicable food legislation for normal foods except where specifically exempted or controlled within that legislation or within specific directives developed to control a PARNUTS category.

The directive also provides for products which fall into PARNUTS categories to be characterised as 'dietetic' or 'dietary' products and prohibits the use of these terms in the labelling, presentation and advertising of foodstuffs for normal consumption. However, it is possible for a foodstuff for normal consumption which is also suitable for a particular nutritional use to indicate such a suitability.

10.2 Categories of PARNUTS products

Directive 89/398/EEC lists, in Annex I to the directive, nine groups of PARNUTS foods for which specific provisions were required to be laid down by specific directives:

1. Infant formulae
2. Follow-on milk and other follow-on foods
3. Baby foods
4. Low-energy and energy-reduced foods intended for weight control
5. Dietary foods for special medical purposes
6. Low-sodium foods, including low-sodium or sodium-free dietary salts
7. Gluten-free foods
8. Foods intended to meet the expenditure of intense muscular effort, especially for sportsmen and women
9. Foods for persons suffering from carbohydrate-metabolism disorders (diabetes).

Article 4(1) of the directive lists the different aspects which the specific directives may cover. These include essential requirements as to the nature or composition of the products controlled by the directive, provisions regarding the quality of raw materials and hygiene requirements, a list of additives that can be used in the products, provisions regarding the labelling, presentation and advertising of the products, any special sampling procedures or methods of analysis required for checking compliance and any permitted deviations from the legislation for normal foodstuffs.

The development of the specific directives proved to be considerably more difficult than originally envisaged by the European Commission and Member State governments when the directive was adopted in 1989. Discussions on the directive for Infant Formulae and Follow-on Formulae preceded the adoption of the framework directive with the original proposal being given in document COM (86) 564 Final in 1986. A revised proposal was adopted after considerable discussion and three further revisions as Directive 91/321/EEC of 14 May 1991.[4] This directive combined the requirements for both infant formulae for infants during the first four to six months of life and follow-on formulae for infants over four months of age. In effect this covered the first group and part of the second group of foods listed in Annex I of the framework directive.

The first draft of a Commission directive on processed cereal-based foods and baby foods for infants and young children was circulated in 1990 as III/9401/90-EN. This directive was adopted on 16 February 1996 as Commission Directive 96/5/EC.[5] Of the other six categories, four have progressed no further than draft directives or early proposals.

The directive on foods intended for use in energy restricted diets for weight control was finally adopted as Commission Directive 96/8/EC in February 1996,[6] nearly seven years after the first draft. The directive on foods for special medical purposes took even longer and was adopted as Directive 99/21/EC.[7]

A proposal for a directive on low-sodium foods and low-sodium or sodium-free dietary salts was circulated by the Commission in 1992 (III/3141/92-EN) but has not been progressed. The work on the directives on gluten-free foods, sports foods and diabetic foods have not progressed beyond initial working parties. Much of the delay was caused by the different needs and priorities within the various Member States, reflecting their concerns about the development of the product categories within their internal market. For example, low-sodium and sodium-free salt substitutes were commonly found on the German market from the mid 1980s onwards, whilst the product category achieved very little penetration in the British grocery market during the same period. Similarly, the market for foods designed specifically for sportsmen and women varied considerably from country to country, both in market size and types of product available.

As a consequence, the Commission notified the meeting of the European Council which was held in Edinburgh on 11 and 12 December 1992 of its intention to reconsider the need for certain legislative initiatives, and in particular the need to continue to develop certain legislation in respect of dietetic foodstuffs. It was agreed at the meeting that the Commission should review whether it was necessary to proceed with the proposed Community legislation for all the groups of products covered by the framework directive. This review was carried out during 1993 and in March 1994 the Commission issued a proposal for a directive to amend Directive 89/398/EEC to reduce the list of categories in Annex I to four, namely:

1. Infant formulae and follow-on formulae
2. Cereal-based foods and baby foods

3. Foods intended for weight control diets
4. Dietary foods for special medical purposes.

The Commission, in an explanatory memorandum to the proposal for the amending directive, outlined the reasons for dropping the five other categories. With regard to low-sodium foods it was noted that the terms such as 'low sodium' and 'reduced sodium' are being used to market both foods for particular nutritional purposes and ordinary foods. The Commission accepted that whilst there was no need for a specific directive on low-sodium foods, there may be a need to define the terms used to describe such products. Definitions and limits have been discussed at Codex Alimentarius (international) level and recommendations have already been made by the Scientific Committee for Food. The Commission is examining possible actions to be taken to define the terms at EU level.

The essential requirement of gluten-free foods is that there should be a virtual (traces only) or complete absence of gluten. This requirement is vital for that proportion of the population suffering from Coeliac disease. As the absence of gluten can be claimed for certain ordinary foods as well as specially designed foods, the Commission felt it was more important to define the conditions under which the term 'gluten-free' can be used and the definition of these conditions would render the requirement for a specific directive unnecessary. The emphasis should be placed on developing a good scientific basis for both the definition and the methods for detection of gluten levels.

The major problem with the development of a directive controlling foods for sportsmen and women was that the products belong to a fast developing and diversified group. Products are also targeted at a wide range of sports needs, such as muscle building, endurance and short intense effort, and the product may emphasise the protein, carbohydrate or micronutrient contents. The Commission believed that the labelling and general requirements of the framework directive could provide sufficient control and information for potential users and that the development and adoption of specific measures were unnecessary. This opinion was later reversed and the category retained in 1999.

The last category, that of foods for persons suffering from carbohydrate metabolism disorders such as diabetes, was also considered not to need a specific directive. The reason given is that people suffering from diabetes and related disorders are usually aware of the particular nutritional needs arising from their specific metabolic problems, and receive advice and guidance on dietary management of them. The Commission considered that the provision of information on the nature of the product and on its nutritional content, as already provided for by the framework directive, was the most important element and that there was no need for a specific directive for this category.

The retention of the requirement for the directive for foods for weight control diets was justified on a number of grounds. The Commission argued that specially formulated products designed to replace one or more of the daily meals and products which can replace the whole of the normal daily diet, are currently being used by a large number of people. As the period of use of the products may vary from days to weeks it was believed that Community rules on the essential composition and labelling of these products were necessary.

A complex group of products are those designed and used for the dietary management of specific physiological or metabolic conditions. There is already a wide and large range of products that have been developed by manufacturers in close collaboration with the medical, dietetic and related professions. These products are used, or should be used, under medical supervision. The progress of scientific and medical developments means that this category of products is in a state of continuous evolution. There is concern that because of the nature of these products their misuse could present certain health risks and a directive was considered necessary to provide for specific labelling requirements for this category of products.

The conclusions of the Committee on the Environment, Public Health and Consumer Protection in July 1995 were that the Committee agreed with the Commission's recommendations not to develop the specific directives on gluten-free foods, low-sodium foods or sports foods. However, the Committee stated that a directive on food for diabetics should be retained in the list of those required.

The indecision on the categories to be included was finally resolved with the adoption of European Parliament and Council Directive 99/41/EC, which gave a revised list of five categories for Annex I.

1. Infant formulae and follow-on formulae.
2. Cereal-based foods and baby foods for infants.
3. Foods intended for use in energy-restricted diets for weight reduction.
4. Dietary foods for special medical purposes.
5. Foods intended to meet the expenditure of intense muscular effort, especially for sportsmen and women.

By the end of 2000, the first four directives had been adopted and work on the sports products directive had only just begun.

10.3 Article 9 of the Directive

The need for procedures for products which are not covered by specific directives was anticipated during the development of the framework directive, and provisions for official notification of these products are laid down in Article 9 of the directive.

Article 9 requires that when a product which does not belong to one of the groups in Annex I of the framework directive is placed on the market for the first time, or where a product is manufactured in a third state, the manufacturer or importer has to notify the competent authority in the Member State in which the product is being marketed.

This notification is initiated by the forwarding of the label, or draft of the label used for the product. In cases where the same product is subsequently placed on the market in another Member State the manufacturer or importer is required to provide the competent authority of that Member State with the same information, together with details of the first notification. A list of the competent authorities in each Member State has been promulgated by the Commission.

The competent authority in each State is empowered to require the manufacturer or importer to produce the scientific evidence to prove the product's compliance with the definition of a food for particular nutritional use as given in Article 1(2) of the directive. Evidence of the particular composition or manufacturing process, which gives the product its particular nutritional characteristics, may also be required. If the scientific evidence for the product is contained in a readily available publication, the manufacturer or importer need only give a reference to the publication.

Whilst the provisions of Article 9 should have applied from 16 May 1991 or from the dates of the implementation of the directive by the individual Member States, very little action has been taken by either the manufacturers or authorities, as it appears to have been generally assumed that although the specific category directives had not been adopted, new products being launched onto the market after May 1989 and which fitted into the general, and somewhat loose, description of the category did not have to be notified.

Article 9(5) of the framework directive required that four years after notification of the directive, the Commission was to send to the Council a report on the implementation of Article 9 together with appropriate recommendations or proposals. This report was in fact delivered in mid-1994 and raised a number of issues. Of the then 12 Member States, six – Belgium, Denmark, Germany, Greece, Ireland and Luxembourg – reported no notifications and the United Kingdom, because of its de-centralised system, had submitted no figures. Within the other five countries there was a wide discrepancy between the number of notifications with the Netherlands reporting only four and Spain 1006.[8]

Of the 1006 notifications reported by the Spanish government, it was subsequently discovered that 673 (67%) were not in conformity with the framework directive 89/398/EEC as the vast majority were for products containing mainly vitamins, minerals, trace elements, amino acids and extracts of plants either alone or in combination. Most of these products would fit into the classification of food supplements which, unless they are designed and marketed for a specific category of people as defined in Article 1(2) of the framework directive, are not considered to be PARNUTS products.

10.4 List of nutrient substances

One of the requirements of the framework directive was still not completed over ten years after its adoption. This is Article 4(2) which calls for the drawing up of a positive list of nutritional substances such as vitamins, mineral salts, amino acids and other substances intended to be used in foodstuffs made and marketed for particular nutritional uses. The list must also contain applicable purity criteria and, where appropriate, conditions under which the substances should be used.

Work on this list was instigated by the need for a list of approved nutrients for the directive on Infant Formulae and Follow-on Formulae (91/321/EEC) and which is included as Annex III to the directive. This list gives the approved sources for the thirteen recognised vitamins together with the co-factors choline and inositol. It also gives the allowed forms of thirteen amino acids, carnitine and taurine. Mineral salts that can be used for the supplementation of seven of the minerals and trace minerals are included in the annex, as these are the only ones covered by the compositional requirements of the formulae as given in Annexes I and II of the directive.

The list for Directive 91/321/EEC was not sufficiently comprehensive to be used as a positive list for all PARNUTS product categories, as it lacked approved sources for a number of trace elements such as selenium, manganese, molybdenum and chromium and also for a number of the amino acids and other nutritional substances.

The Commission originally gave a deadline of 20 May 1993 for the dietetic food industry to submit applications and dossiers for nutrient sources to be added to the list. Almost exactly six years later, on 12 May 1999, the Scientific Committee on Food (SCF) published its opinion on its review of the submissions.[9] Only 24% of the applications were approved by the SCF. On the basis of the SCF recommendations, the European Commission circulated a draft directive on substances which may be added to PARNUTS products. This draft was adopted as Directive 2001/15/EC in February 2001.[10]

Until the list of nutrient substances for all PARNUTS products comes into force, manufacturers still have to comply with a number of national regulations as some countries, such as Germany, Denmark and France, already have positive lists of nutrient substances for dietetic foods, whereas others including Belgium, Ireland, Spain and the United Kingdom do not.

Work on the proposed nutrients list highlighted a serious concern for industry in the area of innovation. Over the past few years there have been a number of significant advances in nutrition and food sciences, and nutrients which were previously not considered, or considered to be unnecessary, have been shown to be more important for the growth and development or maintenance of the human body than originally thought. Under current procedures, amendments to directives can take a considerable time and industry had to discuss with the Commission the need to develop a 'fast track' system for the approval and acceptance of important new substances or processes. A good example of the need for a rapid approval process can be seen in infant nutrition, where recent developments have shown the importance of certain nucleotides and selenium in infant feeds. These do not appear in the list of nutrient sources published as the annex to Directive 91/321/EEC and were included in an amending Directive 96/4/EC five years later.[11]

10.5 Innovative products

During the first part of the 1990s, it became apparent that the framework directive was too rigid and did not easily accommodate innovations in processing and ingredients. After some discussion, European Parliament and Council Directive 96/84/EC was adopted. This amends the framework directive to allow for the introduction into the market place of new PARNUTS products based on technological advances.

To protect consumers, marketing authorisation depends on the Commission consulting with the EU's Scientific Committee on Food. Such authorisation to allow a product to be sold is limited to a two-year period, during which the Commission may require the product to be labelled in such a way as to inform consumers of the different composition of the product from those products authorised under specific directive procedures. During this two-year period, the compositional requirements of relevant directives may be amended to take account of the new product.

10.6 Labelling

Under the PARNUTS framework directive, appropriate PARNUTS products may be labelled as 'dietetic' or 'dietary'. No other foodstuff authorised under the framework food labelling directive (2000/13/EC formerly 79/112/EEC)[12] can use these terms unless it has been authorised through the process allowed under the PARNUTS framework directive. Manufacturers may apply to the Commission and the Standing Committee for Foodstuffs to obtain permission to label foodstuffs as 'suitable' for nutritional/dietary purposes, but any labelling must still follow the procedures set out in the PARNUTS framework directive.

Medicinal claims are prohibited in the PARNUTS framework directive. The labelling and marketing of such foodstuffs must not attribute or imply any properties for the prevention or treatment of disease. A manufacturer may seek an exemption from this ban by applying to the Standing Committee for Foodstuffs with convincing evidence of need.

The labelling of PARNUTS products for which there is no specific directive must include on the label:

- the particular elements of the qualitative and quantitative composition or the special manufacturing process which gives the product its particular nutritional characteristics
- the available energy value in kilojoules and kilocalories and the carbohydrate, protein and fat content per 100 g or 100 ml of the product as marketed and, where appropriate, per specified quantity of the product.

10.7 Directive on Infant Formulae and Follow-on Formulae

Directive 91/321/EEC on Infant Formulae and Follow-on Formulae was the first of the specific directives to be adopted after a number of years of discussions going back originally to the early 1980s. This directive covers infant formulae which are essentially milk substitutes designed for use by infants during the first four to six months of life, and follow-on formulae which are defined as constituting the principal liquid element of the diet for use by infants over four months of age.

The directive is very detailed in its requirements as the Commission and the Scientific Committee for Food considered that products designed for infants required rigorous standards and controls, as at birth some physiological processes involved in the digestion, absorption, hepatic function and renal clearance are not fully developed and that the neonate has a limited tolerance with respect to certain nutrients during its first few months of life. As a consequence, any product developed as the sole source of nourishment for the first few months of the life of the infant has to supply all the nutrients and energy necessary for growth and development of the various organs and tissues. After the age of about four months, the organs are normally sufficiently developed to allow a wider variation in the concentration of nutrients and infants are usually introduced to a wider variety of foods. Follow-on formulae have been developed to act in the transition from breast milk or breast milk substitutes to more solid foods.

The directive is very specific about the composition of these products. They can only be manufactured from protein sources defined in the annexes of the directive and both the quality and the amount of protein to be used is given. The directive provides for the protein to be contributed from either cows' milk sources or from soya protein isolates. Very detailed criteria are given for both. There are limits on both the minimum and maximum amount of fat that must be in the product with specific references to limits for certain fatty acids. The use of sesame oil, cotton seed oil, or fats containing more than 8% trans isomers is prohibited in both groups of products.

Carbohydrate content is also given with maximum and minimum levels and only the carbohydrate sources listed in Annex I to the directive may be used in infant formulae. Minimum levels are given for lactose and maximum for sucrose and pre-cooked or pre-gelatinised starch.

Ten minerals and the thirteen recognised vitamins are controlled in the requirements for infant formulae by both minimum and maximum levels of input. In the follow-on formulae only the minerals iron, zinc and iodine and the vitamins A, D, C and E are controlled. However, it is stated that the concentration of the other minerals should be in amounts at least equal to those normally found in cows' milk and that the ratio of calcium to phosphorus shall not exceed 2.0. Annex III to the directive contains a positive list of nutritional substances (vitamins, mineral sources, amino acids and others) that can be used in the products.

In Article 7, the directive gives very detailed requirements for the labelling of the products including compulsory statements concerning the superiority of breast feeding and that the products be used only on the advice of independent persons having qualifications in medicine or allied professions.

The advertising of infant formulae is controlled by Article 8 of the directive and restricted to publications specialising in baby care and scientific publications. The advertisements may only contain information of a scientific and factual nature and must not imply or create a belief that bottle-feeding is equivalent or superior to breast feeding. The provision of free samples, low priced product or promotional gifts to the general public, pregnant women, mothers or members of their families either directly or indirectly via the health care system or health care workers is not permitted.

The use of educational and information materials on the feeding of infants and young children is also controlled by the directive.

Directive 91/321/EEC only covers infant formulae and follow-on foods for infants in good health, as it was originally intended that products specially designed for infants not in good health should be covered by a separate directive, probably as an off-shoot of the proposed directive on medical foods.

A scientific development which had been under discussion for some time prior to the adoption of the directive was the inclusion of nucleotides and the development of hypoallergenic formulae. These, and other issues on the definition of protein quality and the addition of the trace element selenium, led to the Commission issuing the first draft of an amending directive at the end of 1992. This draft, after a number of changes, was finally adopted as Commission Directive 96/4/EC and modifies a number of the technical details and claims in the original directive. A further amendment was introduced in 1999 giving limits and controls for pesticides.

10.8 Export of infant formulae and follow-on formulae to third countries

A specific directive was adopted by the Council to control the export of infant formulae and follow-on formulae to third countries. Directive 92/52/EEC[13] requires that no product other than infant formulae may be represented as being suitable for satisfying by itself the nutritional requirements of normal healthy infants during the first four to six months of life. In addition, the products exported to third countries must comply with the compositional and labelling requirements laid down in Directive 91/321/EEC on Infant Formulae and Follow-on Formulae, and also with those given in Directive 89/396/EEC on indications identifying the batch or lot, unless otherwise requested or stipulated by the importing country.

The products also have to be labelled in an appropriate language and in such a way as to avoid any risk of confusion between infant formulae and follow-on formulae.

10.9 Directive on Processed Cereal-based Foods and Baby Foods

The second category of foods in Annex I of Directive 89/398/EEC which was to be the subject of its own directive was that covering baby foods. This eventually concentrated on processed cereal-based foods and baby foods for infants and young children. Infants are defined as children under the age of twelve months and young children are those aged between one and three years.

For the purposes of the directive, processed cereal-based foods were divided into four categories:

1. Simple cereals which are or have to be reconstituted with milk or other appropriate nutritious liquids.
2. Cereals with an added high protein food which are, or have to be, reconstituted with water, or other protein-free liquid.
3. Pastas which are to be used after cooking in boiling water or other appropriate liquids.
4. Rusks and biscuits which are to be used either directly or, after pulverisation, with the addition of water, milk or other suitable liquids.

These products are intended for use by infants whilst they are being weaned and by young children for supplementing their diet during the period of adaptation to ordinary foods. The directive also covers baby foods other than the processed cereals.

The composition of cereal-based foods must comply with the requirements of Annex I of the directive and criteria are laid down for cereal content and cereal sources, protein quality and content, carbohydrate content and sources, fat content, limitations on levels of certain fatty acids, calcium and sodium content and the levels of vitamins A and D.

Annex II contains a list of criteria for the composition of baby foods. This includes minimum protein content; the proportions of meat, poultry, fish or offal that need to be in the product with respect to the name of the product; the quantities of total carbohydrate permitted in fruit juices and nectars, fruit only dishes, desserts and puddings; total fat content; total sodium content and levels for vitamin C. Vitamin A levels for vegetable juices are given but vitamin A must not be added to other baby foods and vitamin D must not be added to any baby foods.

A list of nutritional sources that can be added to the products is given in Annex IV. This list includes sources of vitamins, minerals, trace elements, amino acids and a few other substances.

The directive contains a number of controls and requirements for the labelling of the products and the conditions that must be met before certain claims can be made are given in Annex V.

Although the first draft of this directive was issued in early 1991 it was the subject of a number of amendments and adopted as Directive 96/5/EC almost four years after the first draft. The directive was amended in 1999 to add controls and limits for pesticides in baby foods.

10.10 Foods intended for weight control diets

In June 1987, almost two years before the adoption of the framework directive on Foods for Particular Nutritional Uses, a working party of the Scientific Committee for Food published a report of its findings on low-calorie diets (III/1483/EN). The working party made a number of recommendations for these products, particularly in respect of their energy, protein, fat and micronutrient composition. This report was not well received by the industry as it did not reflect the market at the time and also contained unsubstantiated criticisms of the very low calorie products. After further work the Scientific Committee for Food published its official report on 'Food intended for Weight Control Diets' in September 1990. The recommendations from this report formed the basis of the preliminary draft of the directive on Foods Intended for Weight Control Diets (III/3268/91-EN) in early 1991. The draft divided products into three categories:

1. Products presented as a replacement for the whole of the daily diet.
2. Products presented as a replacement for one or more meals of the daily diet.
3. Products presented as an important source of nutrients for persons following a restricted energy diet composed of selected common foodstuffs.

The key point of the draft directive was that following the recommendations of the Scientific Committee for Foods, the energy limits for products in the first category were set at a minimum of 800 kcal (3360 kJ) and a maximum of 1200 kcal (5040 kJ) per daily intake. Meal replacement products in the second category were given a minimum energy content of 275 kcal (1150 kJ) and a maximum of 400 kcal (1680 kJ) per meal. The products in the third category must not exceed 125 kcal (520 kJ). The draft also places limits on protein, fats, fibre, vitamins and minerals.

Later in 1991, a revision to the draft was issued. The main changes were to modify the description of the product category from 'weight control' to 'energy restricted' diets and to introduce a minimum of 10 g of protein per product for the category of products providing an important source of nutrients for people on energy restricted diets. In January 1992 the Commission published an additional annex to the draft which contained an amino acid requirement pattern for the protein. Later in 1992 a second revision to the draft was issued. This excluded very low calorie diets and provided for special rules to be developed for these products at a later date. This draft also removed the third category from the directive and brought the minimum energy content of the second category (meal replacement products) down to 200 kcal (840 kJ).

After the Edinburgh Summit in December 1992, there was little obvious activity on this directive until the Commission made its intentions clear on the future of the specific directives required to be developed under the PARNUTS framework directive. As the directive on products for energy restricted diets was retained in the reduced list of categories, the Commission issued a revised draft directive in March 1994 as Document III/5138/94-EN. The text of this draft was almost identical to the second revision of the previous document (III/3628/91-EN) but with the addition in Annex I of two tables giving vitamin and mineral contents.

The proposals were eventually adopted in 1996 as Commission Directive 96/8/EC on foods intended for use in energy-restricted diets for weight control. The directive came into force fully on 31 March 1999.[14]

The directive covers two categories of product:

1. those products presented as a replacement for the whole of the daily diet and
2. products presented as a replacement for one or more meals of the daily diet.

There are a number of detailed compositional requirements for both categories of product. The energy content of products in the first category must not be less than 800 kcal (3360 kJ) and not more than 1200 kcal (5040 kJ). Meal replacement products in the second category must fall within the range of 200 kcal (840 kJ) to 400 kcal (1680 kJ) per meal. Products in both categories must provide not less than 25% and not more than 50% of the total energy as protein. The quality of the protein is specified, as is the addition of isolated amino acids to improve the nutritional value of the protein if required.

The fat content must not exceed 30% of the total available energy of the product. For products in category 1 (replacement of the whole diet) the linoleic acid content, in the form of glycerides, must not be less than 4.5 g. For meal replacement products the linoleic acid must not be less than 1 g per meal.

For category 1 products the dietary fibre content must be between 10 g and 30 g per daily intake of product.

As an annex to the directive, there is a table listing twelve vitamins and eleven minerals and trace minerals together with amounts for each micronutrient. Products in category 1 must contain at least 100% of the amounts given and those in category 2 at least 30% of each amount with the exception of the amount of potassium per meal which must be at least 500 mg.

The directive also contains a number of labelling requirements specific to the products. There is also a prohibition on the labelling, advertising and presentation of the product containing any reference to the rate or amount of weight loss which may result from the use of the product, or to a reduction in the sense of hunger or an increase in the sense of satiety.

10.11 Foods for special medical purposes

The fourth of the specific category directives to be adopted was that on foods for special medical purposes (FSMP), which after many years of discussion was finally adopted as Commission Directive 99/21/EC.[7] Foods that fall into this category are those which are specially processed or formulated and intended for the dietary management of patients and to be used under medical supervision. The foods are intended for the exclusive or partial feeding of patients with a limited, impaired or disturbed capacity to take, digest, absorb, metabolise or excrete ordinary foodstuffs or certain nutrients contained therein or metabolites or with other medically determined nutrient requirements, whose dietary management cannot be achieved only by the modification of the normal diet, by other foods for particular nutritional uses or by a combination of the two.

This lengthy definition is necessary to encompass the wide range of products of different composition which may differ substantially from each other depending on the specific disease or medical condition of the patients for whom they are intended. In the preparation of the directive the Commission accepted that it was not possible to lay down detailed compositional rules because of the wide diversity of these foods and rapidly evolving scientific knowledge.

The directive classifies foods for special medical purposes into three categories.

1. Nutritionally complete foods with a standard nutrient-balanced formulation which, when used in accordance with the manufacturer's instructions, may constitute the sole source of nourishment for the persons for whom they are intended; they may also be used as a partial replacement or as a supplement to a patient's diet.
2. Nutritionally complete foods with a nutrient-adapted formulation for a disease, disorder or medical condition which, used in accordance with the manufacturer's instructions, may constitute the sole source of nourishment for the persons for whom they are intended; they may also be used as a partial replacement or as a supplement to the patient's diet.
3. Nutritionally incomplete foods with a standard formulation or a nutrient-adapted formulation specific for a disease, disorder or medical condition which are not suitable to be used as the sole source of nourishment.

It is a requirement of the directive that the formulation of a food for a specific medical purpose is based on sound medical and nutritional principles. It is also required that the use of the food in accordance with the manufacturer's instructions is safe, beneficial and effective in meeting the particular nutritional requirements of the patients for whom they are intended. Compliance to these requirements should be demonstrated by generally accepted scientific data.

The directive gives mandatory requirements for the labelling of FSMP that are in addition to those required by the directive on food labelling (2000/13/EC formerly 79/112/EEC). These mandatory requirements include:

- a statement that the product must be used under medical supervision
- a statement whether the product is suitable for use as sole source of nourishment
- where appropriate, a statement that the product is intended for a specific age group
- the available energy value expressed in kJ and kcal
- the average quantity of each vitamin and mineral present in the product
- selectively, the content of nutrients and their components, the declaration of which is appropriate for the use of the product.

It is a requirement that the first two items in the list above are preceded by the words 'important notice' or their equivalent. The labelling must also include the statement:

> For the dietary management of (disease, disorder or condition) relevant to the intended use of the product.

The compositional requirements laid down for foods for special medical purposes are confined to those for micronutrients with vitamins and minerals and their levels of use specified in an annex to the directive.

It is also a requirement of the directive that the manufacturer of an FSMP must notify the competent authority in the member state in which the product is marketed within thirty days of the product being placed on the market. The notification is carried out by forwarding a copy of the label used on the product. There is a derogation that member states need not apply the requirements for notification if they can demonstrate that it is not necessary in order to monitor efficiently the products in their country.

In terms of regulatory compliance, FSMPs must also conform to all the general requirements for PARNUTS products given in Directive 89/398/EEC. For example, any vitamin or mineral source used to supply micronutrients must appear on the permitted list.

The directive on FSMPs also does *not* exempt the products from having to comply with all relevant food legislation in force. This means that ingredients have to be in compliance with the Novel Foods and Novel Food Ingredients Regulation (which also covers novel processes), additives (including sweeteners and colours) must comply with the directives, labelling has to follow all the statutory requirements of food labels, etc.

It is also important to recognise that FSMP's labels and associated literature must be in compliance with Article 6 of Directive 89/398/EEC which states: 'The labelling and labelling methods used, the presentations and the advertising of the product must not attribute properties for the prevention, treatment or cure of human disease to such products or imply such properties'. However, derogations may be provided for by the Standing Committee on Foodstuffs of the EU in exceptional and clearly defined cases.

The prohibition of 'medical' claims does not prevent the dissemination of useful information or recommendations exclusively intended for persons having qualifications in medicine, nutrition or pharmacy.

10.12 Foods intended to meet the expenditure of intense muscular effort, especially for sportsmen and women

In the revised list of specific PARNUTS directives to be developed (Directive 99/41/EC), the fifth directive was that intended to cover products designed and promoted for sports nutrition. A previous attempt to develop this directive ceased in 1992 at an early stage.

In its White Paper on Food Safety, published at the end of 1999, the Commission imposed on itself a timetable for adoption of this directive by December 2001.[15]

The Scientific Committee on Food undertook to produce a report on the composition and specification of foods which would fall into the scope of the proposed directive. This report was published in July 2000 and is expected to form the basis from which the Commission will draw up the draft directive.[16] The SCF report classifies sports nutrition products into four categories:

1. Category A: carbohydrate-rich energy food products
2. Category B: carbohydrate-electrolyte solutions
3. Category C: protein and protein components
4. Category D: supplements and other food components

The scientific rationale and criteria for products in these categories is discussed in the report.

10.13 Food supplements

Food supplements were originally on the list of the PARNUTS categories requiring a specific directive in the annex to the final draft of Directive 89/398/EEC. During the final discussion before adoption of the directive, a number of member states argued that not all food supplements fell into the definition of a PARNUTS product. The category was removed from the directive but, at the same time, the Commission was tasked with developing a specific directive for food supplements.

After a number of false starts, a proposal for a draft directive on food supplements was first circulated by the Commission in March 2000.[17]

After much discussion a revised draft was issued in March 2001.

10.14 References

1. European Council Directive 89/398/EEC. *OJ* of EC L186/27 of 30 June 1989.
2. European Council Directive 96/84/EC. *OJ* of EC L48/20 of 19 February 1997.
3. European Parliament and Council Directive 1999/41/EC. *OJ* of EC L172/38 of 8 July 1999
4. European Commission Directive 91/321/EEC. *OJ* of EC L175/35 of 4 July 1991, amended by Directive 1999/50/EC. *OJ* of EC L139/29 of 2 June 1999.
5. European Commission Directive 96/5/EC. *OJ* of EC L49/17 of 28 February 1996, amended by Directive 1999/39/EC. *OJ* of EC L124/8 of 18 May 1999.

6. European Commission Directive 96/8/EC. *OJ* of EC L55/22 of 26 March 1996.
7. European Commission Directive 1999/21/EC. *OJ* of EC L91/29 of 7 April 1999.
8. Report from the European Commission to the European Council and Parliament 'Implementation of Article 9 of Council Directive 89/398/EEC on the approximation of the laws of the member states relating to foodstuffs intended for particular nutritional uses'. 1994, Brussels.
9. Scientific Committee on Food of the EU 'Opinion on substances for nutritional purposes which have been proposed for use in the manufacture of foods for particular nutritional purposes (PARNUTS)'. SCF/CS/ADD/NUT/20/Final of 12 May 1999, Brussels.
10. European Commission Directive 2001/15/EC. *OJ* of EC L52/19 of 22 February 2001.
11. European Commission Directive 96/4/EC. *OJ* of EC L49/12 of 28 February 1996.
12. European Council Directive 79/112/EEC. *OJ* of EC L33/79 of 8 February 1979, replaced by Directive 2000/13/EC. *OJ* of EC L109/29 of 6 May 2000.
13. European Council Directive 92/52/EEC. *OJ* of EC L179/129 of 1 July 1992.
14. European Commission Directive 96/8/EC. *OJ* of EC L55/22 of 6 March 1996.
15. European Commission 'White Paper on Food Safety'. SANCO/3578/99, 2000, Brussels.
16. Scientific Committee on Food of the EU 'Report on the composition and specification of food intended to meet the expenditure of intense muscular effort, especially for sportsmen'. SCF/CS/NUT/SPORT/5 Final of 11 July 2000, Brussels.
17. European Commission 'Draft Proposal for a European Parliament and Council Directive on Food Supplements'. 14 March 2000, Brussels.

Part III

Case studies

11

Frying oils

R. Fox, Pura Foods Limited, Belvedere

11.1 Introduction: the legal context

11.1.1 The basis of EU law

The law of the European Union on food issues looks satisfyingly similar at every level but the one where it ultimately matters – in the detail. In matters of principle, most countries in Europe legislate in the same way, which ought to be of much comfort to the food technologist. After all, food poisoning bacteria do not respect border patrols, and good nutrition should be the same the world over. Fraud, most countries agree, is a 'bad thing'.

In the national law of most European countries, there exists 'primary legislation', often known as 'laws', and 'secondary legislation' or 'regulations'. The European Union (EU), and the European Economic Community (EEC) before it, overlay that structure with 'directives'. No explanation of food law can go far without defining these terms, and this one will be no exception.

A law is passed by the legislature of a country. The UK's Food Safety Act (1990) sets out the framework for all subsequent legislation. Germany's 'Law on foods and commodities' (*Lebensmittel- und Bedarfsgegendständgesetz* – LMBG) of 1997 does the same, as does Finland's *Elintarvikelaki* (Food Law) 361/1995, and so on. What the primary law does is to grant the executive the right to make regulations. It also imposes duties, creates offences, sets up enforcement agencies and defines penalties. The Food Safety Act says that food shall:

- not be injurious to health
- be of a nature or substance or quality demanded by the purchaser
- not have a false description or misleading label.

So it sets out the principles of safety and fraud.

A regulation (or a decree, an ordinance, an order or a statutory instrument) is made by the government of the country, not its legislature. Who signs it (minister or monarch) varies, but the effect is the same. This is where the food manufacturer must look for hard data. For example, it is the appendices of the Dutch *MVO verordening* 1975 on edible oils and fats that define permitted antioxidants, colours, antifoams and flavours.

The European Council generates directives. Although directives appear to have the force of law, each 'is addressed to the member states', and instructs them to amend their laws if necessary to conform. In most cases, the member states then enact a regulation to implement the EC directive. Nowadays, the EU generates regulations itself, which come into force on specific dates throughout the EU. The national governments then generate a regulation which states hardly more than 'EU Directive number ... enters our law'. Some countries, for example Greece, Spain and Austria, have a food code with the force of law. For definitive information by country, see the tables in sections 11.3 to 11.5, and the sources of information in section 11.7.

11.1.2 The international context: Codex Alimentarius

Codex Alimentarius comprises the consensus of committees from around the world on issues that may be covered by law in some of the member countries. It is a conscious attempt to harmonise regulation across the world. Codex consists of a series of committees, which deliberate on proposals brought by their members and generate standards. Where legislation exists, Codex standards have little force. Nevertheless, governments do take notice of Codex when revising regulations, so it may be seen as a pointer to the future. Since legislation is framed in some countries with an imprecise duty on the manufacturer, Codex can be used in prosecutions, in the absence of strict regulations. The courts then decide how much weight to place on the Codex standards.

With additives having no fixed limit, Codex uses a principle of good manufacturing practice (GMP), implying that the minimum quantity consistent with technological need should be used. European legislation prefers the term *quantum satis* (QS), which means the amount needed, with no implication of minimising usage (see Table 11.1 on page 210).

The Codex Alimentarius Commission has a standard for the composition of vegetable oils,[1] which is the principal reference standard in disputes over the authenticity of oils. Some of the limits are very wide, and more detailed studies taking geographical source into account may be of more use for commercial decisions.

11.1.3 Areas covered by the law

Many countries have compositional standards for certain foods (often staples or those found in past times to have been subject to fraud). The EU has a major preoccupation about olive oil, a commodity closely linked to the economies of several of its members' influential farming constituencies, and one historically subject to systematic attempts at fraud. Thus Regulation 356/92 specifies seven grades of olive oil. Edible oils and fats are subject to such standards in most European countries. In some, there is a distinction between 'seasoning' oils (for salads, sauces, etc.) and those for frying as well. France has a distinction of this nature, revolving on the content of linolenic acid (maximally 2% for frying, although 70% linoleic acid is considered perfectly satisfactory).

Additives are regulated by EU and national law in a consistent and systematic way. The criteria for an additive to be permitted are safety, functionality and need. Safety is obvious, but implies that some poor animal has been tested to destruction to discover the dose at which it dies. Hence the 'need' criterion; legislators do not want to have too many additives tested. Functionality means that the additive has to do something useful. Very few additives are permitted in frying oils, since there are only two agreed functions: antioxidants and antifoaming agents (see section 11.3.5).

Processes and processing aids are controlled in some countries. France is attempting to impose limits on processing aids in edible oil refining at the time of writing. Several countries define the refining operations deemed appropriate for edible oil, and the solvents that can be used.

11.1.4 Non-legal pressures

Most countries have bodies so influential that their pronouncements have almost the force of law. The Food Advisory Committee (FAC) in the UK publishes guidelines that Trading Standards Officers (the local government officers charged with enforcing retail sale laws) attempt to impose on manufacturers as though they were law. One such guideline of interest to oil suppliers is the one discouraging claims on the absence of cholesterol. Austria has in its food code indications of 'an established change in the frying oil' which are guidelines. In Germany, the working group of the Food Chemistry Expert Representatives of the Länder (states) and the BgVV published a position paper on frying fats, which most companies follow as though it was law. This is where the much quoted polar compounds limit comes from. Denmark has a draft order on trans-fatty acids which, though unfinalised owing to lack of scientific consensus, is followed by many manufacturers.

Below this level of influence are respected institutions like the Institut Pasteur in France or commercially powerful groupings like the Institute of Grocery Distribution (IGD) or the British Retail Consortium (BRC), which can effectively impose their guidelines on suppliers. Finally, there are the consumers and no one should be in any doubt of their ability to impose their collective will, especially after the practical disappearance of soya oil in much of Europe because of its genetic modification.

11.2 The structure of the frying industries

11.2.1 The supply chain

The principle of due diligence (see section 11.3) means that each member of the supply chain has to take 'reasonable' precautions to ensure that their suppliers are complying with the law and good practice. The members of the chain are the manufacturers of the foods to be fried, the makers of the frying media, processors who include frying in the preparation of their food, caterers and fast-food outlets, and lastly the retailer of bottled oil. Each has a different interest in the law. An important distinction here is between the nature of the oil as traded, and its condition in the fryer. The user of an oil (caterer, food manufacturer or consumer) is legally protected as to the composition of the oil. He or she also has a right to 'fitness for purpose', so the oil should not deteriorate rapidly when used in accordance with instructions. The user then has the duty not to use the oil beyond its reasonable life, and a different set of rules apply.

Food manufacturers will often have very different requirements from caterers. So, on the one hand, caterers may reuse the oil over a period of many weeks, topping up where necessary. They need a stable long-life oil in order keep costs down, and legislation may well restrict the options there (e.g. on trans-fatty acid levels). As the oil deteriorates, polar compounds, free fatty acids and polymers will build up, and the law often has something to say about that. On the other hand, a manufacturer of pre-fried chips may never have to dispose of the oil, which may have an average residence time of only a few hours. This is because the throughput of chips is so large that even if they take up only 5% of their weight in oil, the contents of the frier are consumed rapidly and need to be constantly replenished. The chip manufacturer can use a relatively unstable but 'healthy' oil such as sunflower. Because the oil never gets old, most of the legislation on end-of-fry life is just not relevant.

11.2.2 Transport

Transport of frying media is covered by all legislation relating to food, but when handled in bulk, there are additional regulations and non-legislative codes of practice. Internationally, a Codex committee is looking into a code of practice for transport. Across the EU, Council Directive 93/43/EEC specifies lorries for bulk road transport. There is a corresponding Commission Directive 96/3/EC covering transport by sea. In the UK, the Seed Crushers and Oil Processors Association (SCOPA)[2] has developed a code of practice covering road tanker construction, tanker registration and identification, operator training, cleaning, and recording of at least the three previous loads.

11.3 The sale of food

The sale of food constitutes a legal act, and a battery of laws applies. 'Sale' usually means offering for sale, or having foods on the premises with the reasonable supposition that they will be offered for sale. Retail sale ('sale to the ultimate consumer') invokes a further set of laws, such as weights and measures and labelling.

11.3.1 Safety and durability

All countries require that food is safe. The expression in Britain is that it should not be 'injurious to health'. Germany has a similar provision, but extends it to any substance that children might mistake for foods. But what does 'safe' mean? Food has a habit of deteriorating into an unsafe condition. Therefore, there is usually a legal requirement for a statement of durability on packaged food. For foods liable to become unsafe, there is a 'use by' date printed on the pack. The 'display until' date is not legally binding, just a guide for the retailer. Foods that deteriorate in quality but not normally into an unsafe condition, have a 'best before' date, which is a legal provision.

Frying oils carry a 'best before' warning, because they contain no water, so do not support microbial growth. Unless stored under very extreme conditions, they do not change chemically in an unsafe way, and even then taste disgusting long before they cause medical conditions. Frying media may be unsafe by virtue of chemicals present before sale. One ought to be able to assume that the manufacturer or trader has done nothing grossly stupid or negligent. However, two examples in 1999 prove that untrue. In Belgium, someone dumped transformer oil into recycled vegetable oil. That was compounded into feed, some of which was fed to pigs, and their fat was rendered into lard. The result was frying lard potentially contaminated with polycyclic biphenyls (PCBs) and dioxins, rather nasty carcinogens.

The second example took place in Indonesia, where palm oil was apparently diluted with cheaper diesel oil. The contamination of around 1% was picked up by a superintendent at Rotterdam, but not before contaminated palm oil had entered the food chain. By chance (or perhaps not – see below), the deodorisation step of physically refining palm oil almost completely removes diesel mineral hydrocarbons. Hydrocarbons are not particularly hazardous to health, and the nauseous smell of diesel may have been the greatest risk to which the consumer could have been exposed.

These examples illustrate that even the most diligent food manufacturers can be caught out by unforeseen events in their supply chain.

11.3.2 The due diligence defence

These examples illustrate well the problem of who is to blame in law for unsafe food. The person or company selling the food to the consumer bears the initial responsibility. Clearly, if sellers commit an unsafe act (for example a butcher contaminating cooked meat with uncooked), they alone probably carry the burden of guilt. If not, they may be able to shift the responsibility back to their supplier, if they have exhibited 'due diligence', and so on back up the supply chain. Due diligence requires that the company in question has taken all 'reasonable' steps to ensure the safety of the product supplied to it. What is reasonable can only be settled in a court of law. A large company will probably be required to take more precautions than a small one. In the case of lard containing PCBs, a legal prosecution would turn on whether the seller might reasonably have foreseen the contamination. The answer is probably no, up to the date upon which the story broke in the press. From that date, any supplier might have been held liable if any lard still in the supply chain reached the consumer containing dangerous levels of PCBs, for they might be expected to have realised the significance of the contaminated animal feed to their supply chain.

In practice, this is what happened, and lard was held up in the supply chain while laboratories worked long hours doing very sophisticated GCMS analyses. They showed that the lard had only 'safe' levels of PCBs.

11.3.3 Safe levels of contaminants

Contaminants such as pesticides, mycotoxins, polycyclic aromatic hydrocarbons (PAHs) and PCBs may be the subject of regulations. Contaminants are controlled at the EU level under Regulation 315/93, with detailed requirements on nitrates, aflatoxin and various metals in subsequent directives. The way in which safe levels are calculated is complex, and depends on the amount of the food typically consumed, and hence the amount of contaminant ingested, compared against a 'no effect' limit in animal studies, usually with a margin of error. The acceptable daily intake (ADI) is the amount of the contaminant judged not to be dangerous, expressed in micrograms of contaminant per kilogram of body weight per day. Given average and peak consumption data, then a maximum recommended limit (MRL) in an individual food can be defined.

In the case of pesticides, a World Health Organisation committee known as JMPR has been establishing MRLs since 1966, and they are the authoritative source. Dioxins and PCBs have been the subject of EU legislative attention since Decision 94/652/EC set up the risk assessment. Apart from in animal tissues (which are prompted by the Belgian dioxin incident), no acceptable levels have been set by the time of writing. The UK's agriculture ministry (MAFF) has published considerable surveillance data in reports on its website, and these can form the basis of judgements on 'safe' levels. The surveillance shows that animal fats are a greater risk to human health than vegetable sources. The consumer seldom understands that such a thing as a no-effect level of a contaminant can exist.

11.3.4 HACCP applied to fried food

The principle of food safety required or recommended in much food safety legislation is that of Hazard Analysis and Critical Control Points (HACCP). A critical control point is a stage in processing which, if not done to specification, compromises the consumer's safety. There are entire books devoted to HACCP, so suffice it to say here that frying will usually be a critical control point, since it may well be the last or even the only heat treatment the food will undergo. In this case, food safety demands that cooking in all parts of the food is sufficient to kill any pathogenic bacteria that might be present. Frying is a high-temperature, short-time cooking method, with heat penetrating from the outside. So the coolest part of the food will always be the centre. Let us consider a few examples:

- Flash frying of fish. Even for double coating of batter with an intermediate frying, the fish remains raw and frozen on the inside. This food is designed for cooking by the consumer. Centre temperature is irrelevant to safety in this case.
- Cooking of bhajees (deep-fried Indian vegetable balls). Here we have an assembled ball, which is then deep fried to cook it. The centre may well have been contaminated by the hands that prepared it, and the ball may be 50 mm in diameter. Clearly, the rate of heat conduction into the centre determines safety, and the bhajee is not safe until the centre has experienced pasteurisation conditions (e.g. at least 72°C for two minutes).
- Frying chips (French fries). Provided the chips have not been mistreated before frying, the centre is essentially uncontaminated, even though the surface may well be. The frying sterilises the surface. In the centre, it is only a matter of cooking the starch to make it taste good.

From these examples, it emerges that measuring the centre temperature of the fried article would be the precaution most likely to ensure safety or eating quality. The food factory may well do so. However, much frying goes on in catering establishments, often by staff with little food safety knowledge. Here, it is better to rely on rigid frying temperatures and times. One usually finds that oil manufacturers print recommended cooking temperatures and times on tins or buckets of oil. That is their contribution to the safety of the food cooked in their product.

11.3.5 Regulation of additives

An additive is an ingredient of a food not normally of itself consumed as a food, and having a function useful in the food. Thus, an antioxidant such as tocopherol is an additive, but salt is not. The law distinguishes a processing aid as an additive which, because of its level or form in the food as sold, has no functionality. Citric acid is used in refining edible oils, and residues remain in the finished oil. It is a processing aid in cooking oil, because it does not perform its normal function in food as an acidity regulator (the function for which it is permitted). As such, it does not need to be declared as an additive.

The legal principle for food additives is that unless they are permitted in the food in question, then they may not be used. The legislation of all countries contains tables of additives listed against various foods. Table 11.1 lists the

Table 11.1 Additives permitted in frying

Additive	Class	Units	Codex Alimentarius CCFAC proposals, 2000	EU
E900 DMPS	Antifoam	mg/kg		10
E310,311,312 gallates	Antioxidant	mg/kg	100	200
E320 BHA	Antioxidant	mg/kg	200	200
E321 BHT	Antioxidant	mg/kg	75	100
319 TBHQ	Antioxidant	mg/kg	Ban (currently 120)	ban
E306,7,8,9 tocopherols	Antioxidant		GMP	QS
E304 ascorbyl esters	Antioxidant	mg/kg	500	QS
Colours in oils & fats generally	Colour			no
E160b annatto in solid fats	Colour	mg/kg	20	10
E160a carotenes in solid fats	Colour	mg/kg	GMP	QS
E100 curcumin in solid fats	Colour	mg/kg	GMP	QS

QS = *quantum satis* (i.e. as much as needed) GMP = good manufacturing practice (i.e. minimum needed)
The appropriate EU legislation is implemented in each of the member states.

permitted additives in cooking oils in Europe. The relevant EC Directive 95/2/EC (as amended) is implemented in all EU countries. For American readers, the notable absentee is tertiary butyl hydro-quinone (TBHQ), which is permitted in the USA and in some other countries. Some palm oil is dosed with TBHQ before shipment, but the refiner then strips it out during refining to comply with local regulation. Even if residues were found, they would be classed as a processing aid, since the concentration would be too low to be technologically significant. Nevertheless, port authorities have been known to take a strict view about the legality of the practice.

The Codex Alimentarius figures given in Table 11.1 are the proposals put forward to the March 2000 meeting of the Codex Committee on Food Additives and Contaminants (CCFAC).[3] There is some doubt about the permitted level of TBHQ. The Codex website quotes 200 ppm, but the Codex standard 19-1981 (revision 2 – 1999) allows only 120 ppm. Any national legislation will of course take precedence, but for the purposes of international trade it would be wise to conform to 120 ppm.

11.3.6 Packaging declarations: nutrition and ingredients

In most countries, there is regulation of what can and must be stated on packaged food for retail sale. In Germany, according to the Food Labelling Ordinance, you must state the following:

- the sales description
- the name of the manufacturer, packer or seller
- a list of ingredients
- the date of minimum durability or the use-by date
- an indication of quantity/amount.

There are similar requirements in other countries, being based on EC directive 79/119 and its amendments.

The list of ingredients has to be in descending order of quantity. A new aspect of labelling is the Quantitative Ingredient Declaration (QUID). Although required by EC Directive 79/119 of 1978, the effective date is as recent as February 2000. At the time of writing, there is some disagreement as to its applicability. A QUID is required if an ingredient is named or implied by the pack graphics, but the directive is strictly not applicable to categories having their own vertical directives. Thus, the countries of the EU have interpreted the directive differently. France has been keen on QUID for some years. Partly, this seems to be because of a consumer preference for multi-component oils, and it is not unusual to find a product claiming it contains 'five oils'. In Britain, by contrast, enforcement officers have been advised to go softly with manufacturers who appear not to be complying with QUID.

One oddity is the requirement to label a flavoured frying product as containing 99% sunflower oil. The remainder is salt, flavour and vitamins. The law states in Britain that packaged food may contain a declaration of nutritional content, but if it does so, then the form is prescribed. I refer the reader to national regulations on this topic.

All the labelling law relates to retail sale. It is intended to inform the consumer. There is no obligation to print any such information on packs for further processing. Business-to-business trade is still governed by any law that uses the word 'sell', but labelling law is typically not one of these. In practice, of course, the buying power of large processors and caterers is such that they require their suppliers' specifications to carry infinitely more information than any retail pack. They may very well be using the specification to work out the declarations on a large number of their own products, which may make a variety of claims, all of which will need to be backed up with specified nutritional and ingredient declarations.

Another point to bear in mind is that the industrial buyer may not have a legal right, but has a due diligence duty to receive accurate and sufficient information from their supplier. A good example is in the labelling of fat components. 'Mono-unsaturates' are defined in labelling regulations as 'fatty acids containing one cis- double bond'. The manufacturer could hide behind a statement of 'total mono-unsaturates', in which they include trans-monoenes. The manufacturer would be unwise to do so, for the customer may transcribe this loose category into the tighter legal definition, and may take their supplier to court if they are prosecuted. The customer would certainly withdraw their business!

Finally, there is the grey area of catering sale. The same design of 20-litre drum of frying oil sold to an industrial customer may be offered in a catering supplies outlet to which the consumer has access for retail purchase. In practice, most manufacturers will play safe and comply with all the legal requirements of retail sale.

11.3.7 Weights and measures

A pack for retail sale must bear an accurate declaration of the weight or volume of its contents. The most frequent form of that declaration is in accordance with average weight legislation. National legislation follows the EC model. Essentially, if a packer opts to use average weight declaration, then:

- the average weight of packs must be not less than the declared weight
- no more than 2.5% of packs can fall more than one 'tolerable negative error'[4] below the declared weight
- effectively none should be more than twice the tolerable negative error below the declaration.

The tolerable negative error (TNE) is defined on a sliding scale, so that it is 9 g between 200 and 300 g, and 1.5% between 1 litre and 10 litres. Packs that comply can be marked with the 'e' symbol. Oils, as liquids, are generally sold by volume. There are no prescribed pack sizes for liquid oils, unlike those for edible solid fats (sold by weight). However, the average weight regulations based on EC Directive 76/211/EEC do specifically mention edible oil in liquid or gel form.

Fats and oils for industrial use lie beyond the scope of prescribed pack size legislation, though still fall within the limits of average weight provisions (which apply up to 20 kg or 20 litres). Regardless of whether the regulations strictly apply to industrial products, the provisions constitute a reasonable guideline for the tolerable negative error; a figure of 150 g should be used between 10 kg and 15 kg, and 150 ml between 10 litres and 15 litres.

11.4 The life of frying oils

11.4.1 Why frying oil has to be discarded

The reader of this book will have gathered that frying oil has to be discarded at the end of its useful life. Intuitively, you can see why; the oil darkens, it thickens, it may contain deposits, and it may acquire an acrid flavour. Fried food will look and taste poor. What is not so clear is why the law should take an interest in frying oil life. Let us examine the arguments.

The law often concerns itself with unfair practices. A frying oil that is used until the food is acrid and blackened could be construed as unfair, inasmuch as the enterprise doing the frying can reduce its costs over a competitor using good practice. Yet it is unusual for the law to intervene when the quality deterioration is so obvious to the consumer. If a food looks and tastes bad, the customer will not buy it. However, poor quality of food is an imprecise legal concept. Therefore some countries, but by no means all, have enacted regulations on specific chemical analyses that measure oil deterioration (see section 11.4.3).

There is only thin evidence that poor oil quality is a health risk. The quantities of the chemicals of concern are enormous by most contamination standards; several per cent for free fatty acids and for polymerised oils, up to a quarter of the oil for total polar compounds. Humans have been consuming these compounds for centuries. Two classes of compound present in abused heated oils have attracted the attention of toxicology researchers: cyclic compounds and polymers. Cyclic compounds arise by cyclisation between C_{15} and either C_{10} or C_{11}, to produce either a six-member or five-member ring respectively. Purified cyclic fatty acid monomers (CFAM) have been isolated from heated linseed oil (a mixture of five- and six-membered rings) and sunflower oil (mainly five-membered rings). It appears that all CFAM are easily absorbed and incorporated into fatty tissues. There have been many metabolic studies. Fatty acids are broken down in cells in two-carbon chunks (β-oxidation), which for CFAM ceases when the ring is encountered. The six-membered rings are excreted rapidly, so have very low toxicity. The five-membered rings are preferentially absorbed, and can lead to decreased liver lipogenesis. The levels of CFAM in these studies was from 0.0075% to 0.15% of diet.[5] In another study,[6] liver enzyme activity was reduced in rats fed purified CFAM derived from used hardened soya oil. The levels fed in both these studies are several fold greater than those normally found in food.[7]

Polymeric compounds are the gums and thickening compounds of used frying oils, and can reach 10% of the oil. They may be neutral or oxidised, and hence polar. Both groups are included in the generic class of 'polar compounds'.[8] In one study,[9] dimers at 0.1%, 1% or 5% of diet were fed to rats. No effect on weight gain was observed. Other studies have used up to 20%, when the rats suffered diarrhoea from what is a highly unbalanced diet.

The above detailed studies are supplemented by many in which used frying oils have been fed. Only mild, if any effects have been reported.[10,11] One extreme study[12] fed rats for eighteen months at 20% of the diet with fresh oil, used oil and its polar or non-polar fractions. Only the diet of 20% polar compounds caused a small reduction in growth and increased liver and kidney weights.

In conclusion, then, the case for legislation is weak on both fraud and safety grounds.

11.4.2 Regulation of fresh frying oils

Ever since oilseed rape was stripped of its high erucic acid content (more than 40% originally), European law has distinguished between the old varieties, which it relegates to industrial use, and edible varieties. The EC Directive 76/621 limits erucic acid to 5% of the fat. All EU countries have incorporated the directive. In practice, no refinery now expects to see more than 1%.

Table 11.2 Regulations on unused frying oils

	EU	Austria	Belgium	Denmark	Finland	France	Germany	Greece	Ireland	Italy	Luxembourg	Netherlands	Portugal	Spain	Sweden	United Kingdom
Erucic acid (max %)	5	5	5	5	5	5	5	5	5	5	5	5	5	5	5	5
Free fatty acids (%)		0.4														
Smoke point (ºC)		205														
Lauric fats		no														
Linoleic acid (max %)						2										
Oils specified														yes		
Used oils														ban		
Trans-fatty acids (max %)				15												

France is somewhat isolated in having a limit on linolenic acid content. The limit could nowadays be argued as a barrier to trade, since there is no safety justification for it, and most European countries have happily used rapeseed oil for deep frying, either in its native form, or hydrogenated to increase its stability. Likewise, much of America fries in soyabean oil, which also exceeds the French 2% linolenic threshold.

Denmark restricts trans-fatty acids in a draft order. The proposed limit is 15% at the time of writing, reducing to 10% eventually. Despite the order not having reached the statute book, many Danish oils and fats businesses already follow the provision.

Austria has regulation of the quality of fresh frying oil in terms of free fatty acids and smoke point. Chapter B30 of the Austrian Food Code goes further in describing as unsuitable: oils with a significant medium or short chain fatty acid content (although it is unclear whether palm oil would be caught by this recommendation); polyunsaturated fatty acids (particularly linoleic acid, and therefore sunflower oil); and animal fats and oils. The United Kingdom, by contrast, uses large amounts of palm oil and lard for frying traditional fish and chips, even though the Draft Regulations and Guidance of Nutritional Standards for School Lunches recommend sunflower oil as the best option if fried food is served in school meals. There is an apparent discrepancy here. Olive oil would typically have a free fatty acid content sufficient to render it ineligible as a frying oil in Austria (see Table 11.2). The important distinction is that while it would be illegal to describe olive oil as a frying oil in Austria, it would not be illegal to use it for deep frying.

Spain's 1983 *Reglamentación Técnico-Sanitaria de aceites vegetales comestibles* (technical and sanitary regulation on edible vegetable oils) adopts a very different approach. It lays down the composition of a whole series of oils, their Lovibond colour scores, iodine values, saponification values, and saturates in the two-position. Only these oils are permitted in frying. Unusually, this vertical regulation also governs the pack sizes and the labelling, subjects that in other countries and in EU directives are covered in horizontal legislation (i.e. covering a single subject for all foods).

11.4.3 End of frying life

Section 11.4.1 argued that the case for regulation of the end of frying life is weak. There is no EU legislation on the subject. I cannot claim that this is because Europe's legislators agree with me. Far more likely is that the Commissioners have seen no need to develop Europe-wide rules on the subject. Much of the EU's legislation is designed to establish a free market across the Union. It is in the nature of frying that its products do not travel well. Fast food cooked in France cannot compete with fast food cooked even in Luxembourg, so the EC has not attempted to force the issue. Food (such as potato crisps) that has been fried legally as part of its processing in one country, can be sold in another country of the EU, even if it would have been illegal to fry in that way in the

Table 11.3 Regulations on end of frying life

	EU	Austria	Belgium	Denmark	Finland	France	Germany	Greece	Ireland	Italy	Luxembourg	Netherlands	Portugal	Spain	Sweden	United Kingdom
Max frying temp. (°C)		180	180			180*				180			180			
Smoke point (min °C)		170					170									
Free fatty acids (max %)			2.5													
Acid value (max)		2.5					2									
Polar compounds (max %)		27				25	24			25			25			
Oxidised fatty acids (max %)		1					0.7									
Dimers and polymers (max %)			25									16				
Viscosity 50°C (max mPa.s)			37													
of liquid oils			27													

* Relates to automatic chip vending machines. 200°C for highly saturated oils.

selling country. The reason is that it would be a restraint of free trade to ban its sale. The only grounds for such a ban are safety issues, and the country where a manufacturer tried to sell the food would have to take legal action to prevent sale. The furore over beef suspected of BSE contamination demonstrates just how much burden of proof is required of a threat to safety.

Individual countries have legislation (or guidelines with the effective force of legislation) on the limit of acceptable quality of frying oil, and hence the point at which it needs to be discarded.

Several countries specify a maximum frying temperature, although France makes an exception in respect of especially stable oils, when 200°C is permitted. Ironically, British manufacturers of long-life oils often recommend a frying temperature of around 190°C, which would be illegal in some European lands. I also find it anomalous that Austria permits frying at 180°C, but sets the smoke point limit at 170°C. It would be quite legal to fry with a continuously smoking oil!

The acidity of the oil can be regulated by acid value (grams of potassium hydroxide per 100 g of oil) or the free fatty acids (grams oleic acid per 100 g of oil). The latter is approximately twice the former.

The consensus when it comes to polar compounds is weak, as befits such a contentious measure. There is also some confusion, because of the varying measures. The data from Hamilton and Perkins[13] for sunflower oil after six minutes of frying illustrate the typical relationship between some of the measures (see Table 11.4). Diglycerides and free fatty acids did not change from the values in the fresh oil.

The most common confusion is between polar compounds and dimers and polymers (the measure in Belgium and the Netherlands). As can be seen, dimers and polymers are only one class of polar compounds (as defined by AOCS method Cd 20-91 or IUPAC method 2.507). The confusion is beginning to be a barrier to free trade in frying oils across the EU, because frying trials in one country are not accepted as evidence in another, solely on the grounds that the correct analytical measure was not used. Germany and Austria have another measure in addition to polar compounds, that of oxidised fatty acids insoluble in petroleum ether.

Table 11.4

	After 6 mins.	Fresh oil
Free fatty acids	0.5%	0.6%
Diglycerides	1.0%	1.2%
Oxidised monomers of triglycerides	3.9%	0.9%
Triglyceride dimers	5.1%	0.6%
Triglyceride polymers	1.1%	0
Total polar compounds	11.7%	3.2%

The various countries cannot agree on an acceptable maximum for polar compounds, Austria quoting 27% and Germany 24%. However, if you are taken to task over an assay of polar compounds, you probably need not worry about these discrepancies: the reproducibility (at 95% confidence) of polar compound analyses across laboratories is quoted as ±2.17% in AOCS method Cd 20-91.

The United Kingdom has no rules on the end of frying life. The absence of criteria in Table 11.3 against other countries should not necessarily be taken as an absence of regulation, and I advise the reader to check with users in the country in question; the rules are not necessarily laid down in formal legislation.

It seems likely that guidelines developed at an international conference at Hagen in April 2000 will have considerable influence in the future (see section 11.6.2).

11.5 Environmental protection

11.5.1 Why the law addresses discharges from frying operations

The principle of all legal systems is that the actions of one person should not damage another person, and any action that causes harm is likely to be forbidden. The environment has recently come to be seen as something that needs protecting in the same way. The earliest environmental legislation worried only about acts that immediately harmed other people. Nowadays, such outcomes as global warming and de-oxygenation of rivers are taken into account.

Frying involves no really nasty materials, no radioactivity or potent chemicals likely to poison or cause cancer. Its hazards are mostly those of excess nutrient. Environmental law covers liquid effluents and waste packaging, which are discussed below. The law usually also deals with smells and gaseous effluents. I shall make no attempt to discuss these, other than to say they are normally based on the principle of nuisance.

11.5.2 Liquid effluents

Fat is the most energy-packed of all foods. Its energy yield in nutrition is 37 kJ/g. In effluent terminology, we refer to its biological and chemical oxygen demand (BOD and COD). These are the weights of oxygen needed to convert the fat to carbon dioxide and water, and hence a measure of the intensity of treatment needed at the sewage treatment works before a relatively clean stream of water can be discharged into a river or the sea. The theoretical COD and BOD of pure fat can be calculated as 3,800,000 ppm. It is this enormous effluent loading that is the principal reason for intercepting fat in fat traps. The other reason is that even a liquid oil will solidify in cold weather (or when it starts thickening in use) and can block drains.

Premises are granted an 'effluent consent' by the authority responsible for such matters. A large food factory will often be expected to pre-treat its effluents to meet the consent. In some cases, the consent applies across a group of factories. A classic case is the big complex of food factories on Grimsby Docks in the UK, where there is a single landlord (the original port authority) and a complex sewer system. Small units, such as a fast-food outlet on a high street, may appear to be less heavily regulated. In practice, assumptions are being made that assess the demands they are placing on the sewage system, without actually measuring them.

Issues such as this are dealt with by departments of government concerned with the environment in most cases, not food authorities.

11.5.3 Waste oils

What is the fate of used oils? They wax up the drains and create an expensive effluent problem if discharged to the sewer. In solid waste, they generate gas problems if sent to landfill. The usual means of disposal is to recycle them via specialist recovery firms.

Spain, Luxembourg and France either implicitly or explicitly ban the use of spent oils in frying. I have been unable to find any other specific ban on the use of spent oil in human foods. The reply from Eire is specific that there is no ban, while Austria and Finland state that it is not covered by food legislation. The Belgian position is typical: 'Article 1 of Royal Order of 3rd January 1975 defines as harmful foods or foodstuffs prepared from raw materials unfit for human consumption'. The implication is that if a spent oil is still within the tolerance of a usable oil, then it can legally be used. In practice, once an oil has been recovered, its traceability is compromised, and good manufacturing practice demands that it should not be used for human consumption. In the light of Belgium's ban on use of recovered edible oil in animal feed (see below), a court might well decide that it was not suitable for human use.

As to spent oils being used in animal feed, my reply from Eire is that 'recovered vegetable oil is not prohibited as an ingredient in animal feeds'. It is allowed in Britain also. Belgium's Adviser General of DGIV has stated that 'a merchant can no longer recycle frying fats to deliver them to an animal feed factory', although a food factory can compound its waste oil into animal feed.

Used frying oils can be used for industrial purposes, although it is probably uneconomic to do so. One likely destination in the future is biodiesel.

The EU is discussing introducing controls on the sourcing, collection, storage and distribution of used edible oil for animal feed. The trigger is the incident at the Belgian Verkest animal feed plant, where transformer oil containing large quantities of dioxins and PCBs found its way into the waste edible oil tanks. The animal feed compounded from this stock caused the death of chickens fed with it, and farm animals across Belgium, Netherlands and parts of Germany had to be removed from the food chain. The bill for lost livestock, recall of food products, not to mention analysis to demonstrate that products were free of the contamination, was colossal. Emergency controls on movement of animals and their products were in place for a year.

11.5.4 Packaging

The packaging materials in contact with foodstuffs are controlled. There is a series of eleven EC directives in place, generally incorporated into national legislation. Most significant of these are those concerned with migration from plastics into the food, and the methods for testing. Since fatty foods are the most at risk of dissolving toxins from plastics, the bottles, pails and plastic cans used for delivering frying oils in packs from 0.5 litre to 20 litres must comply with the directives.

Packaging waste is now controlled, with varying vigour according to the environmental credentials of the national governments. Although the principles are governed by EC Directive 94/62 EC, national schemes vary widely.

11.6 Future trends

11.6.1 The supremacy of EU law

The 'ever closer union' of the states of Europe may be resisted by some governments, and perhaps even by a majority of the population in some nations. Nevertheless, there is a historical inevitability about the convergence of its laws. As far as food law is concerned, it is easy to predict that common features will progressively outnumber differences. Table 11.2 demonstrates that for fresh frying oils there is not much commonality yet. Yet, this is the area where, in my opinion, we are likely to see most progress. The French 2% limit on linolenic acid can be seen as a restraint on trade, especially when so much of Europe already uses rapeseed oil for frying. The Danish trans-fatty acid limit is more difficult to predict, because it could go either way, depending on public opinion and the emerging scientific evidence; either a Europe-wide limit, or dropping of the Danish position.

As I have argued, there is less of a case for legislation on end of frying life, so it would take a major upset for the EU to get involved. Expect the wonderful diversity of legislation to continue.

11.6.2 Pseudo-legislative pressures

Much of the pressure in Europe is in non-legislative but nonetheless effective regulation. The German situation is the most obvious case, with an influential opinion taken by all involved as having quasi-legal status.

An international conference at Hagen[14] in Germany in April 2000 extends the German philosophy of influential opinion. A round-table discussion at the conference led to the internet publication of draft recommendations. They included the following:

- there should be no health concerns associated with consumption of frying fats and oils that have not been abused at normal frying conditions
- analyses of suspect frying fats and oils to confirm abuse should be:

  - total polar compounds (max. 24%)
  - polymeric materials (max. 12%).
- the use of rapid tests correlating with internationally recognised standard tests is recommended.

11.6.3 Codex Alimentarius

Codex teams are working on international standards for additives. Limits for colours annatto, curcumin and β-carotene appear in the latest draft lists. There is a Codex committee working on fat spreads,[15] and another on storage and transport.[16] While Codex has limited legal significance, many legislators take its expert deliberations into account when setting standards. Therefore in some respects Codex is a pointer to the future.

11.6.4 Consumer pressure

In many countries of the EU, consumer pressure goes far beyond legislation. The Unilever submission for novel food clearance on stanol esters was effectively held up by pressure on national governments not to clear 'technological' products. This stems from the shattering experience of the GM issue. American readers may need some explanation.

The European attitude to genetic modification was, for most of the 1990s, benign. The FoodFuture campaign built the case around extensive ethical and consumer reaction surveys, and especially the principle of choice. Innovations in tomato paste and cheese rennet quietly outsold their conventional rivals. The bombshell was the arrival of unsegregated American soya and maize. The press and consumer response was alarm at the extent that these products had permeated the food chain. Most manufactured foods, it seemed, contained soya lecithin, or maize starch, or soya oil, if only as a carrier of minor ingredients and additives.

Consumer pressure came to define 'genetically modified' as 'having an ingredient or additive originating from a genetically modified crop'. Supermarket chains in the UK vied with each other to advertise to their customers that no trace of GM product was present in their own-label food. By contrast, EU Directive 1139/98, and national legislation based on it, defined 'genetically modified' as containing new DNA or protein. Additives were not within its scope, and the legislation envisaged a 'negative list' and a *de minimis* threshold. Subsequently, the EC bowed to the pressure, so that Directive 50/2000 extended the law to additives and flavourings, while 49/2000 established the *de minimis* threshold at 1% of adventitious contamination. This threshold does not allow 1% of GM material, it merely states that for food from a non-GM source, a 1% non-deliberate contamination is not illegal. Some retailers have estimated that this 1% limit translates into 0.05% to 0.1% in finished foods on average. The threshold has legal but little commercial significance; the polymerase chain reaction (PCR) test can detect levels below 0.1%, and any supplier whose products tested PCR-positive within the legal threshold is likely to come under pressure to examine their supply chain.

The consequence for frying is the effective disappearance of soya and maize oils, unless 'identity preserved'. This means that the supplier can show by record-keeping that only crops not genetically modified are included in their product. The British Retail Consortium and Food and Drink Federation have published a standard for how the audit trail can be demonstrated. There are two levels: a minimum compliant standard and a 'recommended best practice' standard, which is exceedingly onerous.

It is interesting to note that the standards of exclusion for GM material are more stringent than in the organic standard. The organic standard was introduced as a thoughtful answer to defining what could be described as organic in selling food. The GM standard is an emotional reaction to a technology that the consumer fears. What will be the next food scare? It would be a brave person who predicted it. However, given that food legislation is driven by the two principles of safety and the prevention of fraud, it is reasonable to expect that any major shift in the law will result from an emerging safety scare.

The long-term concerns with fried food relate to heart disease, and one can see the effect of the link with saturated fat and trans-fats in guidelines at present. Danish draft legislation sets a maximum trans level. The Pasteur Institute in France recommends a target for frying oils of 5% trans maximum.

Here we see health pressures (to reducing saturates and trans) in direct opposition to economic constraints (stability by hydrogenation or using saturated oils). How can the conflict be resolved? The ideal long-life oil is a mono-unsaturated oil laden with natural antioxidants such as tocopherols and other phenols, and with negligible saturated and trans-fatty acids. Rapeseed oil can approach the ideal only by hydrogenation, which builds the *trans* level. Olive oil would be a good candidate if its price could be cut by a factor of six.

The animal feed industry saw a similar issue in the 1970s. Technology in the form of single-cell protein seemed the answer until agriculture cut off its economic legs by developing high-yielding soya. On the principle that nature is always cheaper than chemical industry in the long run, we can look to biotechnology to breed the perfect frying oil. Whether that is by recombinant genetics or by conventional breeding depends on the public climate. The recombinant technology is faster and more specific, but public fears could yet force the conventional route. It is perhaps worth noting that 'conventional' breeding can involve inducing mutations by means that would scare the consumer if he or she knew about them.

11.7 Sources of information

I heartily recommend anyone entering the market in Europe for the first time to check with the relevant national regulatory body, or better still with a company or consultant familiar with the country. The EU law is a good starting point, but significant differences still exist between nation states. There follows a list of contact points with regulatory bodies, publishers of legislation (not normally the same organisation), and Table 11.5 shows the applicable laws and regulations.

My experience is that national authorities appreciate an approach in their own language, but that in many cases it is not necessary. The Scandinavian countries and the Netherlands are very comfortable communicating in English. Germany's Bundesministerium für Gesundheit provided me with a booklet in English on consumer protection in food legislation, in addition to an extensive package in German. Most national government websites have an English version, as well as each of their national languages.

11.7.1 EU and national regulatory bodies

EU

The executive of the European Union consists of the European Commission, organised into twenty-three Directorates General and twelve Services. There are seventeen Commissioners, each responsible for one or more DGs. Depending on the objective of Directives (harmonisation, safety, monetary), it is not possible to generalise about which DG is responsible for food matters, and in any case the national authorities implementing EU law are better points of contact. Each country also has an Office of Representation of the European Commission on its own soil. The European Commission itself is at Rue de la Loi 200, 1049 Brussels, Belgium, tel: +322 299 1111, fax: +322 295 0122, URL: http://www.europa.eu.int.

Austria

Bundeskanzleramt (Bundesministerium für Frauenangelegenheiten und Verbraucherschutz), Gruppe VI/B (Lebensmittelangelegenheiten), Radetzkystraße 2, A-1031 Wien, tel: +43 1 711 72/0, fax: +43 1 713 79 52, DVR: 0000019

Belgium

Ministère Fédéral des Affaires Sociales de la Santé Publique et de l'Environnement, Inspection générale des Denrées alimentaires, Boulevard Pachéco 19, bte 5, B-1010 Bruxelles, tel: +32-2-210 48 43, fax: +32-2-210 48 16, email: ewida@health.fgov.be, URL: http://www.minsoc.fgov.be/en/index.htm or http://belgium.fgov.be/pa.

Denmark

Fødevaredirektoratet, Mørkhøj Bygade 19, 2860 Søborg, Denmark, tel: +45 33 95 60 00, fax: +45 33 95 66 96, email: vfd@vfd.dk, URL: http://www.vfd.dk

Finland

National Food Administration, PO Box 5, 00531 Helsinki, Finland, tel: +358 9 77261, fax: +358 9 7726 7666, URL: http://www.elintarvikevirasto.fi.

France

Direction Générale de la Concurrence, de la Consommation et de la Répression des Fraudes, Bureau D3, 59 boulevard Vincent Auriol, 75703 Paris cedex 13, Tel. (+33) 1 44 97 04 65, Fax. (+33) 1 44 97 05 27

Germany

Bundesministerium für Gesundheit, Am Propsthof 78a, D-53108 Bonn, Germany, tel: +49 228 941 4230, fax: +49 228 941 4989, URL: http://www.bmgesundheit.de/gesetze.

Greece

Higher Chemical Council, Food Directorate, 16 Anast. Tsocha Street, GR-115 21 Ampelokipi, Athens, Greece, tel: +301 64 28 211, fax: +301 64 65 123, telex: 218311 GCSL GR, email: gk-foodiv@ath.forthnet.gr.

Ireland (Eire)

Department of Agriculture and Food, Agriculture House, Kildare Street, Dublin 2, Ireland, tel: +353 1 607 2000, fax: +353 1 661 6263, URL: http://www.irlgov.ie/daff. Food Safety Authority of Ireland, Abbey Court, Lower Abbey Street, Dublin 1, tel: +353 1 672 4711.

Italy

Ministero della Sanità, Dipartimento degli alimenti e nutrizione e della sanità pubblica veterinaria, Piazza Marconi 25, I-00144 Roma, Italy, tel: +39 6 5994 1, fax: +39 6 5994 3676, telex: 613169, URL: http://www.sanita.it/sanita/servizi.htm.

Luxembourg

Ministère de la Santé, 57 boulevard de la Pétrusse, L-2935 Luxembourg, tel: +35 2 478 5527, fax: +35 2 491 337.

Netherlands

Ministerie van Volksgezondheid, Welzijn en Sport, Postbus 20350, 2500 EJ Den Haag, The Netherlands, tel: +31 70 340 6884, fax: +31 70 340 5177, URL: http://www.minvws.nl/international. Ministry of Agriculture Nature Management and Fisheries, Bezuidenhoutsweg 73, Postbus 20401, NL-2500 EK Den Haag, The Netherlands, tel: +31 70 378 4062, URL: http://www.minlnv.nl/international.

Portugal

Ministerio da Agricultura, Instituto da Qualidade Alimentar, Av. Conde Valbom 98, 1100 Lisboa, Portugal, tel: +351 1 796 2161, fax: +351 1 797 1750. Direccão-Geral de Saúde, Divisão de Saúde Ambiental, Ministério de Saúde, Alameda D. Afonso Henriques 45, 1056 Lisboa Codex, Portugal, tel: +351 1 847 5515, fax: +351 1 795 9211.

Spain
Subdirección General de Higiene de los Alimentos, Ministerio de Sanidad y Consumo, Paseo del Prado 18, 28071 Madrid, Spain, tel: +34 91 596 1000 /596 1608, fax: +34 91 596 1547 /596 1548. Ministerio de Agricultura Pesca y Alimentación, Paseo Infanta Isabel 1, 28014 Madrid, Spain, tel: +34 91 347 5403, fax: +34 91 347 5006

Sweden
Statens livsmedelsverks, Box 622, S-75126 Uppsala, Sweden, tel: +46 18 17 5500, fax: +46 18 10 5848, email: livsmedelsverket@slv.se, URL: http://www.slv.se.

United Kingdom
Ministry of Agriculture, Fisheries and Food, Ergon House, 17 Smith Square, London SW1P 3JR, tel: +44 20 7238 3000, fax: +44 20 7238 6591, email: consumer@info.maff.gov.uk, URL: http://www.maff.gov.uk. Food Standards Agency, PO Box 31037, Ergon House, 17 Smith Square, London SW1P 3WG, tel: +44 20 7238 6480, fax +44 20 7238 6763, email: helpline@foodstandards.gsi.gov.uk, URL: http://www.foodstandards.gov.uk.

Codex Alimentarius Commission
Vialle delle Terme di Caracalla, 00100 Roma, Italy, tel: +39 06 57051, fax: +39 06 5705 4593, email: codex@fao.org, URL: http://www.codexalimentarius.net.

11.7.2 Publishers of legislation

EU
The *Official Journal of the European Communities* is the official source of all EU directives, decisions and regulations. Copies of relevant issues are available in each of the relevant community languages through the official seller in each community state. This is mostly the seller of national legislation documents. It is also now published on the internet at http://www.europa.eu.int/eur-lex/en/oj/index.html. Each issue is accessible for twenty days following the date of publication.

Austria
Official journal, *Bundesgesetzblatt der Republik Österreich*, from Österreichische Staatsdruckerei, Rennweg 16, A-1037 Wien, Austria, tel: +43 179 789294, fax: +43 179 789419, available on the Internet to subscribers: http://www.verlagoesterreich.at/gbbl/. Austrian Food Code published by Brüder Hollinek (projektsitz), Luisenstrasse 20, 3002 Purkersdorf, Austria, tel/fax: +43 223 167 365.

Belgium
Moniteur Belge (in French, or *Belgisch Staatsblad* in Flemish), la Direction du Moniteur Belge, rue de Louvain 40-42, 1000 Bruxelles, Belgium, tel: +32 2 552 2211, fax: +32 2 511 0184, URL: http://www.moniteur.be for issues since June 1997.

Denmark
Decrees published by: Schultz Information, Herstedveg 10-12, 2620 Albertslund, Denmark, tel: +45 43 63 23 00

Finland
Suomen Säädöskoskoelma (in Finnish, or *Finlands Författningssamling* in Swedish) from: OY Edita AB, FIN-00043 Edita, Finland, journals freely available on http://www.edita.fi/fs.

France
Journal Officiel de la République Française, from: Journaux Officiels, Service Information Diffusion, rue Desaix 26, 75727 Paris Cedex 15, France, tel: +33 1 40 58 79 79, fax: +33 1 45 79 17 84, URL for issues since January 1998: http://www.journal-officiel.gouv.fr.

Germany
Bundesgesetzblatt, Bundesanzeiger. Verlagsgesellschaft mbH, Südstraße 119, 53175 Bonn, tel: +49 228 382 080, fax: +49 228 382 0836; Bundesanzeiger, tel: +49 221 976 680, fax: +49 221 976 68115, URL: http://www.bundesanzeiger.de.

Greece
Government gazette available from the National Printing Press, fax: +30 1 523 4312. Greek Food Code published by GS Alysandratos and Associates, Colokoltroni 13, 15772 A Ilisia, Greece, tel: +30 1 775 6767, fax: +30 1 959 2322.

Ireland (Eire)
The Government Publications Office, Molesworth Street, Dublin 2, tel: +353 1 671 0309.

Italy
Official gazette published by: Istituto Poligranco e Zecca della Stato, Direzione Editonale, Settore vendite e abbonamenti, Via Marciana Marina, 00199 Roma, Italy, tel: +39 06 8508 2307, fax: +39 06 8508 4117.

Luxembourg
Mémorial Journal Officiel du Grand-Duché de Luxembourg, Imprimérie de la Cour Victor Buck, BP 1341, Luxembourg 1013, tel: +35 24 99 86 61, fax: +35 24 99 41 64.

Netherlands
PBO publications from: SER, Bezuidenhoutsweg 60, postbus 90405, 2509 LK Den Haag, Netherlands, tel: +31 70 3 499 499, fax: +31 70 3 832 535. *Nederlandse Staatscourant* and *Staatsblad van het Koninkrijk der Nederlanden* available from tel: +31 70 3 789 880, fax: +31 70 3 789 783, email: adv.staatscourant@sdu.nl.

Portugal
URL for electronic version of the official journal: Diário da República Electrónico http://www.dr.incm.pt, email: dre@incm.pt.

Spain
Boletín Oficial published by: La Librería del BOE, Trafalgar 27, 28010 Madrid, Spain, tel: +34 91 538 2121, email: clients@com.boa.es.

Sweden
Ordinances and guidelines published by the Statens livsmedelsverk (see section 11.7.1 above).

United Kingdom
All legislation published by: Stationery Office, PO Box 276, London SW8 5DT.

11.7.3 Legislation: the laws and regulations

In Table 11.5, the numbers are reference numbers to the list below, which is sorted by country. The absence of an entry does not necessarily mean that no legislation exists, merely that I, and those I have consulted, have not identified a specific law or regulation covering that topic.

11.7.4 General references

1. Codex Alimentarius Commission Alinorm 99/17 Appendix II, see Codex website at *http://www.fao.org/es/esn/codex*
2. Seed Crushers and Oil Processors Association, 6 Catherine Street, London WC2B 5JJ, tel: +44 20 7836 2460, fax: +44 20 7379 5735
3. Codex Committee on Food Additives and Contaminants: Codex General Standard for Food Additives available on *ftp://ftp.fao.org/codex/ccfac32/fa9915be.pdf.*
4. The 'tolerable negative error' is defined in legislation for various pack sizes, falling as a percentage of the declared size as that size increases. For example, for 500 g packs it is 15 g (3%), and for 15 l it is 150 ml (1%).
5. IWAOKA W T, PERKINS E G: Metabolism and lipogenic effects of the cyclic monomers of linolenate in the rat. *JAOCS* **55**, 734–8 (1978)

(References continued on page 232.)

Table 11.5 The laws and regulations by country

	EU	Austria	Belgium	Denmark	Finland	France	Germany	Greece	Ireland	Italy	Luxembourg	Netherlands	Portugal	Spain	Sweden	United Kingdom
Primary law on food safety and avoidance of fraud		A1	B1	DK1	SF1	F1	D1	G1	EI1	I1	L1	NL1	no	E1	S1	UK1
Composition of oils and fats		A2	B2			F2	D2	G2			L2	NL2	P1	E2	no	no
Erucic acid content	EU1	A3	B3	DK2			D3	G3	EI2	I2	L3				S2	UK2
Permitted colours	EU2	A4	B4	DK3	SF2	F3	D4	G4	EI3	I3	L4	NL3	P2	E3	S4	UK3
Permitted flavours	EU3	A5	B5				D4	G5	EI4	I4	L5	NL4	P3	E4	S3	UK4
Permitted miscellaneous additives (e.g. antioxidants)	EU4	A6	B6	DK3	SF3	F3	D4	G6	EI5	I3	L6	NL5	P4	E5	S4	UK5
Labelling	EU5	A7	B7	DK4	SF4	F4	D5	G7	EI6	I5	L7	NL6	P5	E6	S5	UK6
Weights and measures in general		A8				F5		G8		I6	L8		P6	E7		UK7
Average weights	EU6								EI7		L9				S6	UK8
Control of used edible oil			B8	no		F6			no		L10			E8		no
Packaging recycling or disposal	EU7								EI8	I7	L11	NL7	P7	E9	S7	UK9

Key: An entry 'no' means that controls do not exist. The absence of an entry should not be taken to imply the same.

European Union
EU1 erucic: Directive 76/161
EU2 colours: Directive 94/36
EU3 flavours: Directive 88/388, completed by Directive 91/71
EU4 miscellaneous additives: Directive 95/2
EU5 labelling: Directive 79/112, amended by 93/102, 95/42
EU6 average weights: Directive 76/211, amended by 78/891
EU7 packaging: Directive 94/62

Austria
A1 safety: Lebensmittelgesetz 1975, BGBI nr. 86
A2 oils: Österreichische Lebensmittelbuch, 3. Auflage, Kapitel B30, section 1.6
A3 erucic: Erucasäureverordnung BGBI nr. 468/1994
A4 colours: Farbstoffverordnung BGBI nr. 541/1996
A5 flavours: Aromenverordnung BGBI nr. 42/1998
A6 miscellaneous additives: Zusatzstoffverordnung BGBI II nr. 383/

1998
A7 labelling: Lebensmittelkennzeichnungsverordnung, 1993
A8 weights: Fertigpackungsverordnung BGBI nr. 867/1993, BGBI nr. 132/1995, BGBI II nr. 139/1997

Belgium
B1 safety: Loi du 24/1/77
B2 oils: Arrêté royal du 23/4/74 (edible oils), arrêté royal du 22/1/88, amended 3/5/99 (frying), arrêté royal du 2/10/80 (human consumption)
B3 erucic: Arrêté royal du 26/2/76
B4 colours: Arrêté royal du 9/10/96
B5 flavours: Arrêté royal du 24/1/90
B6 miscellaneous additives: Arrêté royal du 1/3/98
B7 labelling: Arrêté royal du 13/11/86
B8 used oils: opinion of Advisor General, DG4 Agribex, Brussels of 7/2/2000.

Denmark
DK1 safety: Fødevareloven nr. 471 af 1/2/98
DK2 erucic: Bekendtgørelse nr. 57 af 22/1.99
DK3 additives: Bekendtgørelse nr. 942 af 1/6/97,
DK4 labelling: Bekendtgørelse nr. 598 af 14/8/93 (general), 198 af 20/3/92 (nutrition)

Finland
SF1 safety: Elintarvikelaki (Food Law) 36[illegible]/1995, 1/4/95
SF2 colours: ruling 1756 of 1/1/96
SF3 additives including emulsifiers and antioxidants: ruling 811/1997 0f 2/8/99
SF4 labelling: regulation 794/1991 of 10/5/91 and ruling 795/1991 of 1/6/91

France
F1 safety: Code de la Consommation (loi no. 93-949 of 26/7/93)
F2 oils: décret du 11/3/1908, décret no. 73-139 du 12/2/73, arrêté du 19/11/90
F3 additives: arrêté du 2/10/97
F4 labelling: arrêté du 7/12/84, décret no. 03-1130 du 27/9/93, arrêté du 3/12/93
F5 weights: arrêté du 21/3/85
F6 used oil: loi du 15/7/75

Germany
D1 safety: Lebensmittel- und Bedarfsgegendständegesetz (LMBG 1/1/97)
D2 oils: Guidelines on edible oils and fats of 17/4/97
D3 erucic: Verordnung vom 24/5/77 (BGBl I p782), last amended 26/10/82 (BGBl I p1945)
D4 additives (including colours, flavours, others): Verordnung vom 29/1/98 (Bundesgesetzblatt 1998 Teil 1, nr. 8)
D5 labelling: Lebensmittel-Kennzeichnungsverordnung vom 6/9/84 (BGBl I p1221), last amended by BGBl I p460; Nährwert-Kennzeichnungsverordnung vom 25/11/94 (BGBl I p3526)

Greece
G1 safety: Greek Food Code (GFC)
G2 oils: articles 70 to 78 of GFC
G3 erucic: articles 70 to 78 of GFC
G4 colours: article 33 of GFC
G5 flavours: article 44 of GFC
G6 miscellaneous additives: article 35 of GFC
G7 labelling: article 11of GFC
G8 weights: EC directives have been implemented

Table 3.5 Continued

Ireland

EI1 safety and fraud: Sale of Food and Drugs Act 1879, amended 1879 and 1899. See also Health Acts 1947 (no.28), 1953 (no.26), 1970 (no.1) and Statutory Instrument (SI) no.333 (1991)

EI2 erucic: Health (Erucic Acid in Food) Regulations 1978 (SI no.123), amended by SI no. 67 (1992) and SI no. 271 (1982)

EI3 colours: SI no. 344 of 1995

E14 flavours: SI no. 22 of 1992

E15 miscellaneous additives: SI no. 128 of 1997

EI6 labelling: SI no. 205 of 1982 as amended

EI7 average weight: Packaged Goods (Quantity Control) Act 1980 (no.11), Regulation SI no.39 of 1981, as amended by Metrology Act 1996 (no.27)

EI8 Waste Management Act 1996 (no. 10), Regulations SI no. 242 of 1997

Italy

I1 safety: Law no. 283 of 30/4/62

I2 erucic: Law no. 659 of 9/10/80

I3 all additives: Ministerial decree no. 209 of 27/2/96

I4 flavours: Decree no. 107 of 25/1/92

I5 labelling: Legislative decree no. 109 of 27/1/92 (labelling); Legislative decree no.77 of 16/2/93 (nutrition)

I6 weights: Law no. 690 of 25/10/78; Presidential decree no.391 of 26/5/80; Decree-law no. 450 of 3/7/76 as amended

I7 packaging: Legislative decree no. 22 of 5/2/97

Luxembourg

L1 safety: Act of 25/9/53 as amended

L2 oils: Regulation of 4/8/75

L3 erucic: Regulation of 29/12/77

L4 colours: Regulation of 19/3/97

L5 flavours: Regulation of 20/12/90

L6 miscellaneous additives: Regulation of 10/4/97

L7 labelling: Regulation of 16/4/92 as amended (labelling); Regulation of 22/6/92 (nutrition)

L8 weights: Regulation of 26/11/81 as amended

L9 averafe weights: Regulation of 19/10/77 as amended

L10 used oil: Ministerial order of 30/6/99

L11 packaging: Regulation of 31/10/98

Netherlands

NL1 safety: Warenwet (Dutch Commodities Act); Dutch Food Law of 10/12/91

NL2 oils: Decree of 10/4/75

NL3 colours: Decree of 27/9/95

NL4 flavours: Decree of 24/1/80

NL5 miscellaneous additives: Decree of 23/9/96

NL6 labelling: Decree of 10/12/91 as amended (labelling); Decree of 7/9/93 (nutrition)

NL7 packaging: Regulation of 30/12/97

Portugal

safety: no basic food law

P1 oils: Decree-law no. 32/94 of 5/2/94; Order no. 928/98 of 23/10/98

erucic: EC directive applies

P2 colours: Order no. 759/96 of 26/12/96

P3 flavours: Order no. 620/90 of 3/8/90

P4 miscellaneous additives: Decree-law no. 363/98 of 19/11/98

P5 labelling: Decree-law no.560/99 of 18/12/99 (labelling); Order no.

751/93 of 23/8/03 (nutrition)
P6 weights: Order no. 359 of 7/6/94; Order no. 1198/91 of 18/12/91
P7 packaging: Decree-law no. 366-A/97 of 20/12/97 and Order no. 29-B/98 of 15/1/98

Spain
E1 safety: Spanish Food Code, as approved by decree no.2484/1967 of 21/9/67 as amended
E2 oils: Royal decree no. 1011/1981 of 10/4/81 as amended; Royal decree no.308/1983 of 25/1/83 as amended
erucic: EC restrictions apply
E3 colours: Royal decree no. 2001/1995 of 7/12/95
E4 flavours: Royal decree no. 1477 of 2/11/90 as amended
E5 miscellaneous additives: Royal decree no. 145/1977 of 31/12/97
E6 labelling: Royal decree no.1334/1999 of 31/7/99 (labelling); Royal decree no. 930/1992 of 17/7/92 (nutrition)
E7 weights: Royal decree no.723 of 24/6/88; Royal decree no. 1472 of 1/12/89
E8 used oil: Order of 26/1/89 as amended
E9 packaging: Law no. 11/1997 of 24/4/97

Sweden
S1 safety: Food Act (SFS 1971:511 as amended)
oils: none
S2 erucic: Ordinance SLV FS 1993:15
S3 flavours: Ordinance SLV FS 1996:1 as amended
S4 additives including colours: Ordinance SLV FS 1999:22
S5 labelling: Ordinance SLV FS 1993:19 as amended (labelling); Ordinance SLV FS 1993:21 (nutrition)
S6 average weights: Ordinance STAFS 1993:18
S7 packaging: Ordinance SFS 1994:1235 as amended

United Kingdom (England & Wales – different statutory instrument numbers relate to Scotland and to Northern Ireland)
UK1 safety: Food Safety Act 1990
UK2 erucic: 1977/691, amended by 1982/264
UK3 colours: 1995/3124
UK4 flavours: 1992/1971, amended by 1994/1486 and 1996/1499
UK5 miscellaneous additives: 1995/3187
UK6 labelling: 1996/1499
UK7 weights: Weights and Measures Act 1963, Order 1988/2040, amended by 1990/1550, 1994/2868
UK8 average weights: 1986/2049, amended by 1987/1538, 1992/1580, 1994/1852
UK9 packaging: 1997/648

6. LAMBONI C, SEBEDIO JL, PERKINS EG: Cyclic fatty acid monomers from dietary heated fats affect rat enzyme liver activity. *Lipids* **33**, 675–91 (1998).
7. MAHUNGU S M, ARTZ W E, PERKINS E G: Oxidation products and metabolic processes. Chapter 2 in *Frying of Food* ed. Boskou D and Elmadfa I, Technomic (1999).
8. KOCHHAR S P (ed.) *New Developments in Industrial Frying*, SCI, London (1997).
9. PERKINS E G, TAUBOLD R: Nutritional and metabolic studies of noncyclic dimeric fatty acid methyl esters in the rat. *JAOCS* **55**, 632–4 (1978).
10. KEANE K W, JACOBSON G A, KRIEGER G H: Biological and chemical studies on commercial frying oils. *J Nutr* **68**, 57–74 (1959).
11. MARQUEZ-RUIZ G, DOBARGANES M C: Nutritional and physiological effects of used frying fats. In *Deep Frying, Chemistry, Nutrition and Practical Applications*, ed. Perkins E G, Erickson M D, AOCS (1996).
12. BILLEK G, GUHR G, WAIBEL J: Quality assessment of used frying oils: a comparison of four methods. *JAOCS* **55**, 728–33 (1978).
13. HAMILTON R J, PERKINS E G: Chemistry of Deep Fat Frying, in *New Developments in Industrial Frying*, ed. Kochar S P SCI, London (1997).
14. Third International Symposium on Deep Fat Frying, Hagen/Westphalia (Germany) 20–21 March, 2000, organised by Deutsche Gesellschaft für Fettwissenschaft e.V., conclusions to be found on http://www.gdch.de/dgf/recomm.htm, reported in *Inform* **11**, 630–1 (2000).
15. Codex draft standard for fat spreads and blended spreads, Alinorm 99/17, Appendix VI.
16. Codex recommended code of practice for the storage and transport of edible oils and fats in bulk, Alinorm 99/37, para 165 and Appendix VII.

12

Functional foods

P. Berry Ottaway, Consultant, Berry Ottaway and Associates Limited, Hereford

12.1 Introduction

Food law always lags behind innovation and developments, sometimes by more than a decade. This was particularly true in the late 1990s with advances in nutritional science and the general acceptance that some aspects of foods can contribute to health in other ways than by an adequate supply of classical nutrients.

From a relatively slow start, the concept of a functional food has been gaining ground world-wide, and has also been attracting the attention of the major multinational food companies. Within Europe there has been increasing recognition of functional foods by the national authorities, particularly in the area of health claims for foods.

The composition and proposed marketing of many functional foods can introduce a number of anomalies in the application of current EU food legislation and the following is a case study of the proposed introduction of such a product.

12.2 Product description

The product, which was in an advanced stage of development in a country outside the EU, was also being considered for the European market. The concept of the product was a powdered beverage mix which could be made up with milk, water or fruit juices and which provided not only protein, carbohydrate and fat, but also a wide range of micronutrients, added fibre sources and fructo-oligosaccharides. The fructo-oligosaccharides were added for their prebiotic benefits.

The micronutrients included all those listed in the directive on nutrition labelling (90/496/EEC)[1] plus a number of other trace minerals such as selenium, copper and manganese. The antioxidant vitamins C, E and beta-carotene were present at levels of daily intake above the recommended daily allowances (RDAs) used for nutrition labelling purposes in Europe. In addition the formulators of the product wanted to add other carotenoids such as lycopene and lutein as additional *in vivo* antioxidants.

The product was to be available in three flavours with the appropriate colours. The marketing objective in the country of origin was to market the product not only as a nutritious beverage but also to position it for convalescents, athletes and as a meal replacement for weight control purposes.

12.3 Product positioning in the European market

The definition of the product from the marketing point of view was found to be critical. Some of the recommended uses fell into the definition of dietetic as given in the directive on foods for particular nutritional uses (89/398/EEC)[2] known as the PARNUTS directive. There is a specific directive in force, 96/8/EC,[3] which controls both the composition and labelling of foods marketed as meal replacements for use in weight control diets. The composition of such meal replacements must comply with very detailed criteria with respect to the energy, protein, carbohydrate, fat and micronutrient content. The product as developed did not meet all the criteria so the decision had to be made to market it as a convenient healthy beverage applicable to a range of lifestyles.

12.4 Product composition

A detailed investigation had to be carried out on every component, whether ingredient or additive, to ensure compliance with the various European laws.

12.4.1 Protein

The protein contribution was made up of both isolated soya protein and casein (milk protein). The specifications and origins of both had to be checked. To comply with the Council Regulation (EC) No. 1139/98 on the labelling of genetically modified soya or maize,[4] the provenance of the soya had to be traced and certification obtained that it did not contain genetically modified protein or DNA.

There is a European Directive 83/417/EEC laying down the specification and quality criteria for caseinates and the ingredient had to be checked for compliance.[5]

12.4.2 Fat

The fat contribution was supplied from an oil high in polyunsaturated fatty acids plus some lecithin. The specification and typical analyses of the oil were obtained to ensure that permitted maximum levels of erucic acid in the oil were not likely to be exceeded. Erucic acid is a normal constituent of seed oils which has been shown to have detrimental effects on health if consumed in large quantities. There is a limit for erucic acid from oils used in compounded foods where the overall fat content of the food exceeds 5%. The details are given in Directive 76/621/EEC,[6] with the method of analysis in Directive 80/891/EEC.[7]

Directive 76/621/EEC also gives a derogation for member states to apply the provisions of the directive to foods where the total fat content is equal to or less than 5%.

Due to the high polyunsaturated content of the oil it was more susceptible to oxidation (rancidity) than many oils used in such products. The presence of a number of mineral salts in the product also increased the risk of rapid rancidity. Permitted antioxidants for fats and oils are given in the directive on additives other than colours and sweeteners (95/2/EC).[8] As the proposed source of the oil was North America, discussions had to be conducted with the suppliers to ensure that the oil was adequately protected using only the antioxidants and permitted levels given in the directive. As European law in this area differs significantly from that in the USA, this caused considerable problems which were only resolved by changing to a different grade of oil to that originally preferred.

Legal complications were also encountered with the lecithin. In early 1999, when the work was being carried out, Regulation (EC) No. 1139/98 on the labelling of genetically modified (GM) soya and maize was in force but the proposed exclusion list for highly processed soya and maize derivatives had not been adopted. The original source of lecithin proved positive when tested for DNA from GM soya. Alternative supplies were offered from South America but these were found to have differences in their functional characteristics from the original source. The legal requirement to label the lecithin containing the GM DNA was based on its primary function in the product. At that time, the regulation only covered food ingredients from GM soya and maize and specifically excluded additives and flavourings. Lecithins are approved additives and appear in Annex 1 to Directive 95/2/EC as being generally permitted in foodstuffs.

According to the formulator of the product, the lecithin had been included for two reasons: the first was technological, to improve wetting-out characteristics when the powder was mixed into the liquid; and the second was nutritional, to provide a source of phospholipids. This situation, where substances can have dual roles in foods as additives and nutrients, is not uncommon in European food law. The decision was originally made in early 1999 that the primary function of lecithin in the product was as a technological additive and the marketing department had the option not to make the label statement on the GM source. This option was negated in January 2000 when the European Regulation (EC) 50/2000 on the labelling of additives and flavourings from GM sources was adopted.[9] However, while the decision was made to specify the lecithin as an additive, its contribution had to be added to the total fat content given in the nutrition information on the label. Directive 90/496/EEC on nutrition labelling specifically includes phospholipids in the definition of fat.

12.4.3 Carbohydrate

The main carbohydrate component of the product consisted of a mixture of dextrose, fructose and maltodextrin. As these ingredients can be produced from maize, the GM status of each had to be determined. The product was also found to contain relatively small amounts of sorbitol, principally as a component of some compounded ingredients. Under European law the definition of carbohydrates for labelling purposes includes the polyols, of which sorbitol is one, but requires the energy calculation for the contribution from sorbitol to be made with a different factor. Carbohydrates (excluding polyols) must be calculated on the basis of 4 kcal/g whereas polyols are at 2.4 kcal/g. Also, for the purposes of nutrition labelling, the statement of carbohydrate content had to be subdivided into sugars, polyols and starch. The legal definition of sugars includes all monosaccharides and disaccharides in the foods, but excludes polyols.

12.4.4 Fibre

The added fibre and fructo-oligosaccharides presented a number of legal problems, particularly in the quantification of the fibre content. There is no formal definition of dietary fibre in European food law. When Directive 90/496/EEC was adopted, the definition of fibre was given as follows:

> fibre means the material to be defined in accordance with the procedure laid down in Article 10 (of Directive 90/496/EEC) and measured by the method of analysis to be determined in accordance with that procedure.

This statement resulted from a major disagreement in 1990 among the member states as to what constituted dietary fibre. Almost ten years later this had not been resolved. The debate has revolved around the specific components of non-digestible plant matter that collectively contribute to the dietary fibre content and should be included in the analysis. An early definition of dietary fibre was 'the plant polysaccharides and lignin which are resistant to hydrolysis by the digestive enzymes of man'. There are more recent schools of thought that fibre should be defined more closely as 'non-starch polysaccharides (NSP)'. NSPs are the major fraction of fibre and are chemically identifiable. The British authorities are in favour of the definition being NSP as measured by the Englyst method, while some other member states prefer to use a concept of fibre that includes other substances such as lignin. There are a number of ways of chemically determining the fibre content of foods depending on the definition used and the components included in the definition.

The British authorities have consistently insisted that the quantitative declaration of fibre content given on the label should be based on NSP measured by the Englyst method. Since the adoption of the directive on nutrition labelling in 1990, they have been persistent in trying to persuade the other member states to agree to accept a definition based on NSP. This has not succeeded and in 1996 the British Ministry of Agriculture, Fisheries and Food issued a statement saying that although they still regarded the Englyst method, or one giving the same results, as the official method in UK law, manufacturers could label with fibre content determined by other methods such as the AOAC, provided the method of analysis was declared on the label.[10] However, it was also stated that claims for fibre could only be made on values determined by the Englyst method. This meant that in order to make a claim for a high fibre content for the product in the UK, it had to contain at least 6 g per 100 g of NSP measured by the Englyst method. If the AOAC method, commonly accepted by other EU countries, had been used, the actual declaration of fibre content on the label would have been much greater, but no claim for it could be made on the pack or in advertising. In 1999 the British government issued proposals for the labelling of fibre to be based on the AOAC method. Unfortunately, as a result of a long consultation period, these proposals had not been introduced into law by the time the product was launched.

A further complexity surrounding the fibre claim was that the product contained significant amounts of inulin and other fructo-oligosaccharides which have been shown to have a beneficial (prebiotic) effect on the gut microflora. The status of these substances in terms of labelling have been the subject of considerable debate in Europe, particularly in the UK. In 1997 the British government Committee on the Medical Aspects of Food Policy (the COMA Committee) was asked to consider the inclusion of inulin and oligofructose in the UK definition of dietary fibre for labelling purposes. In a statement in April 1998 the committee reported that it had agreed to retain the existing definition based on NSPs, but to consider additional categories for declaring resistant starch and non-digestible oligosaccharides on product labels. A year later the committee was asked to reconsider its decision and reported again in April 1999 that, having reviewed submissions in support of the application, the committee was not convinced by the evidence presented and concluded that inulin and oligofructose should not be included in the definition.[11] This decision meant that these substances could not be added to the fibre content for the UK label and had to be declared separately on the statement of nutritional information. These differences in the definition of fibre meant that there would have to be a dichotomy in the marketing strategy between the UK and the rest of Europe.

12.4.5 Micronutrients

The proposed addition of a wide range of micronutrients caused considerable problems. The fortification of foods with vitamins and minerals is one aspect of European food law that is still under discussion and has not yet reached the stage of a draft directive. The complexities of legislating in this area were outlined in a European Commission discussion paper in 1997.[12]

Across Europe there has been no consistent approach to the addition of vitamins and minerals to foods. Some countries, such as the UK and the Netherlands, have relatively liberal policies while others, such as Spain and Ireland, impose very strict controls. Throughout the EU, there are fifteen different sets of laws covering the addition of vitamins and minerals to foods and they all differ in detail from each other. Some of these differences are very significant.

Many countries use the RDA as a basis for the legal control of vitamins or minerals. For example, Germany permits vitamins, excluding vitamins A and D, to be added to foods up to 3 × RDA per daily intake of the food. The addition of vitamins A and D to foods is prohibited, with only some specific exceptions. Belgium has complex legislation with vitamins A and D at 1 × RDA, the B vitamins, C and E at 3 × RDA and most others at 2 × RDA, and formal notification is required. Spain and Ireland both impose an upper limit of 1 × RDA on micronutrients in foods. In addition, French law does not permit the addition of micronutrients to foods but permits restoration of vitamins at levels between 80% and 200% and minerals at between 80% and 120% of the natural content of the food before processing, although the addition of micronutrients to foods for particular nutritional uses (PARNUTS products) is allowed in France. Some countries only permit micronutrient addition to foods on the basis of individual product authorisations. A further complication was that some of the countries only allowed the addition to foods of the twelve vitamins and six minerals that appear in the Annex to Directive 90/496/EEC. This list does not include the trace elements copper, manganese and selenium which appear in the specific PARNUTS directives and are recognised as being essential in these foods.

These complexities meant that the added micronutrient content of the product either had to be reduced to the lowest common denominator from each country's requirements or a marketing decision had to be made to market the product initially in those countries where the legislation was most compatible with the original product concept. Eventually, the latter option was preferred as the first substantially reduced both the number of micronutrients that could be added to the products and also their level of input.

Once the list of micronutrients had been determined, the chemical forms in which they could be added also had to be checked. This was important as, for the minerals particularly, some of the salts and organic forms of the nutrients are not officially recognised in many countries in Europe. The only guideline that was available was the draft list and opinion of the European Commission's Scientific Committee on Food relating to approved nutrient sources for PARNUTS foods.[13] This draft was published in May 1999, almost exactly ten years after the requirement for such a list was given in Article 4(2) of the PARNUTS Directive 89/398/EEC.

12.4.6 Novel foods and novel ingredients

The original proposal for the product included a number of substances, such as the carotenoids lutein and lycopene, that were to be added for their *in vivo* antioxidant functions. The list of those to be considered consisted of a number of plant extracts including some with high levels of polyphenols.

The first task was to check each proposed substance on the list for acceptability, both in terms of their status with regard to the Council Regulation (EC) No. 258/97 on novel foods and novel ingredients,[14] and to the national situation in the countries of intended sale.

Although the Regulation 258/97 had been in force for over two years at the time the review was carried out, the situation was found to be very confused. The main criterion for classification as a novel food or ingredient is that the substance had not previously been used for human consumption in the European Community *to a significant degree*. Unfortunately, no formal definition of the phrase 'to a significant degree' had been agreed between the European Commission and the fifteen member states. Interpretations varied from that which accepted evidence that the substance had been on sale in a food product in one member state before 15 May 1997 (the date that the regulation came into force), to evidence of a large distribution and sales in more than one member state.

The main problem encountered was that although evidence of prior sale in the UK and the Netherlands could be found for some of the substances, enquiries determined that they were not considered acceptable for use in food products in other countries such as Germany and France. The investigations highlighted a major weakness in the system. The intention of Regulation 258/97 is that novel foods and novel ingredients are reviewed and approved by the competent authority in the member state of intended first sale. This is carried out with the knowledge of the other fourteen member states who are notified of the application by the European Commission. Once approved, the substance should be accepted throughout the EU. No provision was made in the regulation or in any other European food legislation for mutual recognition of foods and ingredients that have been introduced into one of the national markets a few years before the regulation came into force. This has left a number of ingredients, including some on the proposed list for the product, in a situation where approval for use still has to be obtained on a country-by-country basis.

The lutein and lycopene were an anomaly. Both carotenoids have been approved for use as food colourings in Europe, but with restricted levels of input. They both appear in Annex V of Directive 94/36/EC on colours for use in foodstuffs and their permitted use is restricted to specified categories of food and drink.[15] The category that most closely defined the product was 'non-alcoholic flavoured drinks' and the maximum level given for lutein and lycopene was 100 mg/l either individually or in combination. The calculation is based on the pure dye content of the colour. Therefore, if both carotenoids were used in compliance with the directive, the maximum allowable level of each would be 50 mg/l of the ready-to-consume drink. There is no official recognition of either lutein or lycopene in any other area of European food law. In terms of the proposed formulation, an acceptable level of both carotenoids could be achieved within the limits given in the directive on colours but it was noted that, in order to comply with the legislation in some countries, these ingredients would have to be listed as colours in the declaration of ingredients on the label.

12.4.7 Colours and flavours

Both food colours and food flavourings are controlled by European directives. The proposed colours and the levels of use had to comply with Directive 94/36/EC. As many of the proposed colours were carried on a base or were in the form of a lake, details of the pure dye content of each had to be obtained to enable the appropriate calculations to be made.

The situation with flavourings was that the flavouring components and source materials used in their production came under the requirements of Directive 88/388/EEC as amended.[16] This directive includes a list of substances that are considered undesirable from the point of view of human health and are therefore restricted.

As most food flavourings are compounded proprietary mixtures, certification had to be obtained from each of the proposed suppliers that their flavouring complied with the requirements of the directive.

12.5 Functional claims

One of the important aspects of the product concept was that both nutrition and health claims could be made for the product. This is a very difficult legal area and the problems have been highlighted by recent developments in the functional food market.

European food legislation in the form of the directive on food labelling 79/112/EEC[17] specifically prohibits the attribution to any foodstuff of the property of preventing, treating or curing a human disease, or any reference to such properties. In this context, human disease has been interpreted as any ailment, injury or adverse condition, whether in body or mind. Under EU Directive 65/65/EEC,[18] the medicines directive, any claim expressed or implied that a product can prevent, treat or cure a disease or condition is regarded as a medicinal claim and the product has to be treated in law as a medicine. Similar legislation applies in most countries of the world.

As there is often a very fine line dividing medical claims and health claims, much rests on the semantics and presentation. A statement that folic acid helps prevent neural tube defects in the foetus would be considered a medical claim if made for a food or food supplement. By saying that folic acid helps with the development of a healthy nervous system in the foetus, the claim becomes a health claim. Both claims are scientifically correct, but the first relates to the prevention of an adverse condition and can only be made for an authorised medicine.

Since 1980 there have been a number of abortive attempts by the European Commission to introduce pan-European legislation on health claims. After a series of fruitless meetings through the early 1990s with no agreement reached by the then smaller number of member states, the European Commission abandoned its plans for a European directive on claims. This meant that the individual national regulations have continued to remain in force, resulting in a diversity of approaches across the EU.

In the absence of pan-European legislation on claims, there has been considerable activity in a number of member states of the EU. For complex legal reasons these developments have been either in the production of Codes of Practice or agreements between the food industry and the national regulatory authorities. Codes of Practice on health claims have been introduced into Sweden,[19] Belgium,[20] the Netherlands[21] and the UK.[22]

In Spain, a voluntary agreement on health claims has been signed between the Spanish food industry and the Ministry of Health.[23] In France, the French Conseil National de l'Alimentation (National Dietary Council) has been considering evidence from consumer groups, scientists and industry and in 1998 drafted an opinion and proposals on claims linking diet and health.[24] While all six countries have taken slightly different approaches to obtain the same objectives, all the codes and agreements agree on the major points.

There is general consensus that some provision should be made to allow health claims for foods and that these should be in addition to the currently permitted nutrient function claims (e.g. calcium is required for healthy bones and teeth).

There is also general recognition that health claims could be made for foods containing substances other than the traditionally recognised nutrients. A good example of such a case is where the cholesterol-lowering effects of a fat-based spread with the brand name Benecol are attributed to plant stanol esters, which are, at present, not recognised as nutrients.

The codes or agreements require that the claims are substantiated by appropriate scientific evidence. With the exception of the Belgian code, the others require a review and acceptance of the scientific evidence in support of the claim to be carried out by a panel of independent experts.

The most detailed requirements are given in the British code on health claims. This requires that the claim must be based on a systematic review of all the available scientific evidence relating to the validity of the claim, including published scientific literature. The conclusions of the review must be based on the totality of the evidence and not just that which supports the claim. The evidence must also be based on the most methodologically sound human studies and not just biochemical, cellular or animal studies, although other sources of information such as epidemiological evidence and animal, biochemical or cellular studies should be used to support the substantiation.

The evidence must be able to demonstrate that the food will contribute to a positive and significant physiological benefit when consumed by the target population as part of their normal diet. The claimed effect must be achievable with the consumption of a reasonable amount of the food on a regular basis or by the food making a reasonable contribution to the diet.

Expressed in the British and Dutch codes and implicit in the others is that the food as presented to the consumer must be demonstrated to produce the desired effects. The use of surrogate studies or reliance on bibliographical evidence only would not be acceptable.

It is also important that it can be demonstrated that the claimed effect is maintained over a reasonable period of time and is not just a short-term response to which the body later adjusts. The exceptions allowed are for health claims that are for situations that are only relevant for a short- or medium-term benefit. A good example is the requirement for folic acid pre-conceptually and for the first twelve weeks of pregnancy in the case of neural tube defects.

These stringent requirements for health claims resulted in considerable discussion between the development and marketing teams of the company as it became apparent that further studies would be required to support the substantiation. A cost–benefit appraisal had to be carried out by the marketing department to see if the cost of acquiring the extra data was likely to be returned from additional sales if the claim was made.

12.6 Packaging

The proposed packaging had not only to be tested for the barrier properties in terms of product stability but also checked for compliance with a number of laws.

The first group of legislation that had to be checked was that dealing with materials and articles in contact with food. Council Directive 89/109/EEC[25] as amended is the framework directive which lays down a general requirement that all materials that come into contact with food should not transfer their constituents to food in quantities that could endanger human health or make the food unacceptable to the consumer. The directive also restricts the use of vinyl chloride monomer in the manufacture of food-grade plastics and places controls on the use of regenerated cellulose film coming into contact with food.

Under the framework directive there are a number of more specific directives, including Commission Directive 97/48/EC[26] on plastic materials and articles in contact with food and directives on the methods of testing the migration of the constituents of plastics to foods.

As the inner surface of the packaging that came into contact with the product was a plastic, these directives were particularly relevant and certification of compliance to the directives had to be obtained from the supplier of the packaging.

Although not directly part of European food law, the requirements of the directive on packaging and packaging waste (94/62/EC) also had to be considered.[27]

Aspects of this directive have a direct relevance to the packaging of the product. The main ones to be considered were the requirements that the packaging used must be the minimal subject to the safety, hygiene and acceptance for the packed product and for the consumer. The packaging used must be recoverable through at least one of the following:

- material recycling
- incineration with energy recovery
- composting or biodegradation.

The directive also permits packaging to be reusable, but this was not appropriate for the product concept. Any noxious or hazardous substances in the packaging must be minimised in any emissions, ash or leachate from either incineration or landfill.

Within the directive there is also a very specific requirement for heavy metal limits in packaging or any of its components. These limits, which refer to the total concentration of cadmium, mercury, lead and hexavalent chromium, refer to packaging in general and not just to that which comes into contact with food. The heavy metal limits were 250 parts per million (ppm) by weight for any packaging used on or after 30 June 1999 and these reduce to 100 ppm by weight on or after 30 June 2001. Again, assurances had to be obtained from the manufacturers of all the components of the packaging that their products complied with the directive, both in terms of recovery and the ability to meet the heavy metal limits.

Instructions also had to be given to the packaging designers to ensure that the requirements for minimalisation of the packaging were taken into consideration.

12.7 Labelling

Once the pack design had been agreed it was important that all the legal requirements could appear on the label in the appropriate manner. The list of compulsory requirements is given in Directive 79/112/EEC (as amended) and the main ones include the name of the product as a generic name, the list of ingredients, instructions for use, a statement of minimum durability, storage conditions and the name of the manufacturer, packer or seller established within the EU. The declaration of minimum durability, in this case a 'Best before end:' statement and the storage conditions, were based on the results of the product shelf-life trials.

In the case of the declaration of ingredients, the marketing department had a preference to exercise the option of declaring the additives by their generic names as given in the directives instead of using the 'E' numbers for the additives.

As the original development of the product had taken place in North America, many of the values in the nutrition labelling had to be adjusted to the European requirements given in Directive 90/496/EEC. Not only were the factors for calculating the energy content from the energy nutrients different between the two continents but there were also significant differences in the calculation of the activity of a number of vitamins. For example, the thiamin (vitamin B_1) level had been calculated originally on the basis of input of the salt, whereas in Europe the declaration is as the amount of the thiamin cation present. There was also a major discrepancy in the calculation of vitamin A activity in betacarotene.

The formulation was checked to ensure that the composition did not trigger any statutory warnings or statements such as those required in Directive 94/35/EC on sweeteners where the presence in a product of the intense sweetener aspartame or polyols require prescribed warning statements. The national requirements in this area also had to be taken into consideration. For example, in the UK there is a voluntary agreement between the British Department of Health and the food industry that products containing added vitamin A (as retinol) should carry a warning for pregnant women if the vitamin A content of the recommended daily intake of the product exceeds 800 μg. The contribution of betacarotene to the vitamin A content is excluded from this requirement.

12.8 Manufacture

The manufacture of the product was a dry-blending process followed by the spraying into the mix of oil and lecithin. While it was envisaged that the production for the launch of the product would be carried out in North America, there was a requirement to find a suitable production facility in Europe.

As part of the evaluation of potential contract manufacturers, a technical, quality and hygiene audit was carried out on the main contenders. The hygiene part of the audit was designed to ensure that all the requirements of Directive 93/43/EEC[28] on food hygiene were in compliance. This included confirmation that a hazard analysis and critical control point assessment (HACCP) had been carried out by the company as required by the directive.

12.9 References

1 European Council Directive 90/496/EEC on nutrition labelling for foodstuffs. *O.J. of E.C.* L276/40 of 6 October 1990.
2 European Council Directive 89/398/EEC on foodstuffs for particular nutritional uses. *O.J. of E.C.* L186/27 of 30 June 1989.
3 Commission Directive 96/8/EC on foods intended for use in energy-restricted diets for weight control. *O.J. of E.C.* L55/22 of 6 March 1996.
4 European Council Regulation (EC) No. 1139/98 on compulsory labelling of certain foodstuffs produced from genetically modified organisms. *O.J. of E.C.* L159/4 of 3 June 1998 as amended by Commission Regulation (EC) No. 49/2000. *O.J. of E.C.* L6/13 of 11 January 2000.

5 European Council Directive 83/417/EEC relating to caseins and caseinates. *O.J. of E.C.* L237/25 of 26 August 1983.
6 European Council Directive 76/621/EEC on the maximum level of erucic acid in oils, fats and foodstuffs for human consumption. *O.J. of E.C.* L202/35 of 28 July 1976.
7 European Council Directive 80/891/EEC on method of analysis for determining the erucic acid content of fats and oils. *O.J. of E.C.* L254/35 of 27 September1980.
8 European Parliament and Council Directive 95/2/EC on food additives other than colours and sweeteners. *O.J. of E.C.* L61/1 of 18 March 1995.
9 European Commission Regulation (EC) No. 50/2000 on the labelling of foodstuffs containing additives or flavourings that have been genetically modified. *O.J. of E.C.* L6/15 of 11 January 2000.
10 Ministry of Agriculture, Fisheries and Food (UK). Definition and Determination of Dietary Fibre. Letter and statement dated 1 May 1996.
11 Food Safety Information Bulletin, MAFF, UK. Definition of Dietary Fibre. Bulletin No. 109, June 1999.
12 European Commission DG III. 'Addition of vitamins and minerals to foods and food supplements – a discussion paper'. III/5934/97, 1997, Brussels.
13 Scientific Committee on Food of the European Commission. Opinion on substances for nutritional purposes which have been proposed for use in the manufacture of foods for particular nutritional uses. SCF/CS/ADD/NUT/20/Final of 12 May 1999.
14 European Parliament and Council Regulation (EC) No. 258/97 concerning novel foods and novel food ingredients. *O.J. of E.C.* L43/1 of 14 February.
15 European Parliament and European Council Directive 94/36/EC on colours for use in foodstuffs. *O.J. of E.C.* L237/13 of 10 September 1994.
16 European Council Directive 88/388/EEC on flavourings used in foodstuffs and source materials used in their production. *O.J. of E.C.* L184/61 of 15 July 1988.
17 European Council Directive 79/112/EEC relating to the labelling, presentation and advertising of foodstuffs. *O.J. of E.C.* L33/1 of 8 February 1979.
18 European Council Directive 65/65/EEC relating to proprietary medicinal products. *O.J. of E.C.* 22 February 1965.
19 Federation of Swedish Food Industries *et al.* Health Claims in the Labelling and Marketing of Food Products. Food Industry's Rules (Self-Regulating Programme). Revised programme of 28 August 1996.
20 Federatie Voedingsindustrie/Fédération de l'Industrie Alimentaire (Belgium). Health Claims Code of Conduct, draft, 21 October 1998.
21 Voedingscentrum (Netherlands). Code of practice assessing the scientific evidence for health benefits stated in health claims on food and drink products. April 1998.
22 Joint Health Claims Initiative (United Kingdom). Code of Practice on

Health Claims on Foods. Final text, 9 November 1998.

23 Joint 'Ministerio de Sonidad y Consumo' (Ministry of Health, Spain) and Federación de Industrias de Alimentación y Bebidas (Spanish Federation of Food and Drink Manufacturers) agreement on health claims on foods of 20 March 1998.

24 Conseil National de l'Alimentation (France). Allégations faisant un lien entre alimentation et santé. Avis No. 21, 30 June 1998, Paris.

25 European Council Directive 89/109/EEC relating to materials and articles intended to come into contact with foodstuffs. *O.J. of E.C.* L40/30 of 11 February 1989.

26 European Commission Directive 97/48/EC on plastic materials and articles in contact with food (*O.J. of E.C.* L222 of 12 August 1997) amending Council Directive 82/711/EEC. *O.J. of E.C.* L297/26 of 23 October 1982.

27 European Parliament and Council Directive 94/62/EC on packaging and packaging waste. *O.J. of E.C.* L365/10 of 31 December 1994.

28 European Council Directive 93/43/EEC on the hygiene of foodstuffs *O.J. of E.C.* L175/1 of 19 July 1993.

Index